Issued under the authority of the H
(Fire and Emergency Planning Dir

Fire Service Manual

Volume 1
Fire Service Technology, Equipment and Media

Physics and Chemistry for Firefighters

HM Fire Service Inspectorate Publications Section
London: The Stationery Office

© Crown Copyright 1998
Published with the permission of the Home Office
on behalf of the Controller of Her Majesty's Stationery Office

Applications for reproduction should be made in
writing to The Copyright Unit, Her Majesty's Stationery Office,
St. Clements House, 2–16 Colegate, Norwich, NR3 1BQ

ISBN 0 11 341182 0

Cover and half-title page photograph: Hampshire Fire and Rescue Service

Printed in the United Kingdom for The Stationery Office
J43472 4/98 C50

Physics and Chemistry for Firefighters

Preface

In order to understand how fires behave and how they can be extinguished, it is necessary to understand some physics and chemistry. This book is divided into three parts which will introduce the reader to the relevant physical and chemical processes and then show how these work together in the phenomenon that we call fire.

Extinguishing fires is a matter of interrupting one or more of these processes so that burning cannot continue. Firefighters will appreciate that it is, therefore, important to acquire a good understanding of what happens in a fire, in order to be able to choose the best method available to extinguish it, and to avoid making it worse.

In the first part of this book, some of the physical properties of matter will be discussed. Some materials are heavier than others, bulk for bulk. Some heat up more easily than others. These and other properties greatly affect the way that materials behave when they are involved in a fire.

In the second part, the chemical processes relevant to fire will be discussed. Besides the burning process itself, the way that materials behave chemically in fire will also be discussed. It is hoped that firefighters will gain an understanding of the dangers that new materials present and the way the weapons they have to fight them work.

The third part of the book discusses fire extinction.

This book replaces: The Manual of Firemanship, Book 1 – Elements of combustion and extinction.

The Home Office is indebted to all those who assisted in the production of this book, in particular, Edinburgh University Department of Civil and Environmental Engineering, Dr. John Brenton and Dr. Dougal Drysdale.

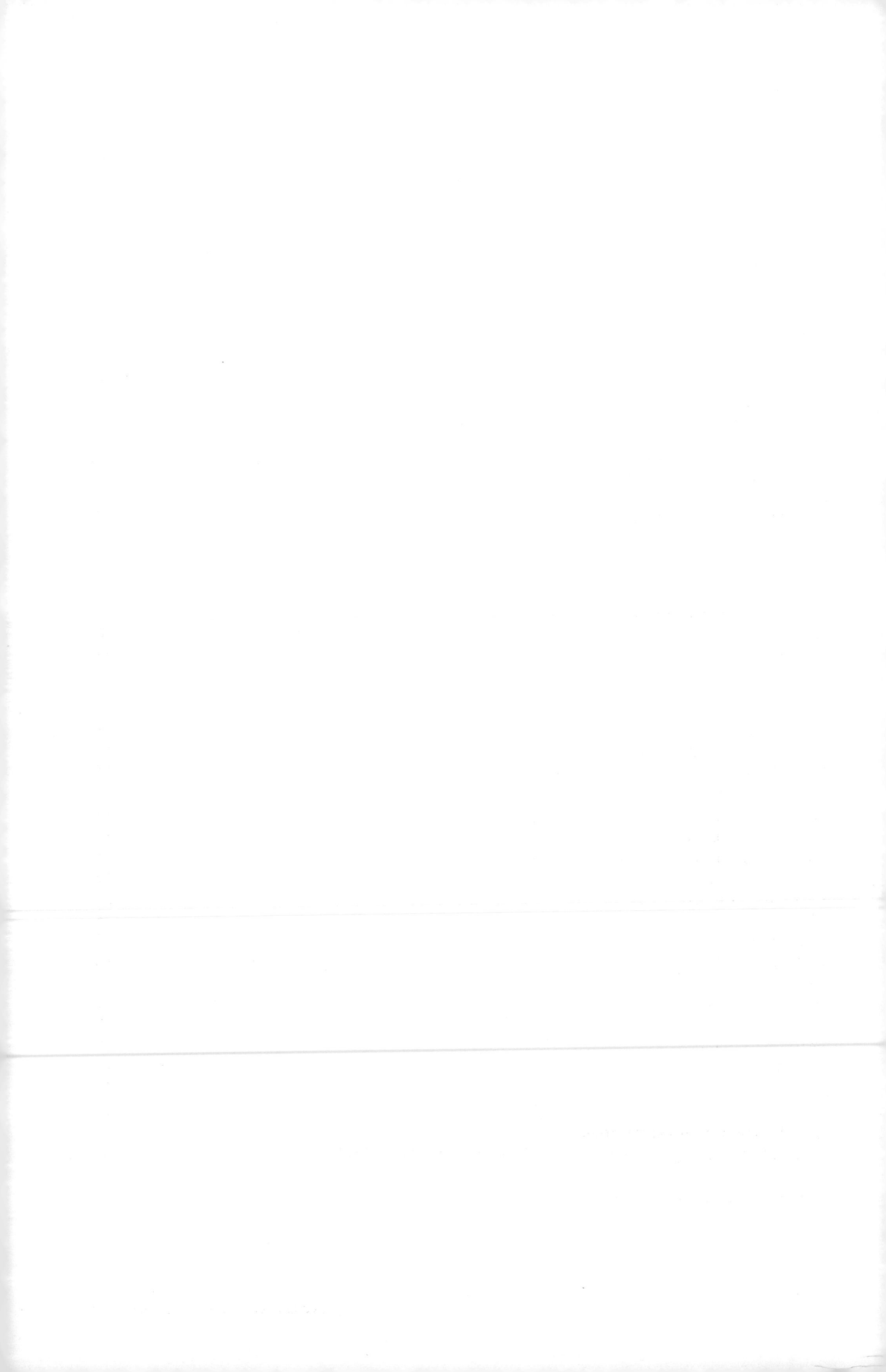

Physics and Chemistry for Firefighters

Contents

Preface	iii
Chapter 1 Physical properties of matter	1
Density	
1.1 Vapour density	2
1.2 Liquids of different density	2
1.3 Gases of different density	3
1.4 Matter and energy	5
1.5 Melting, boiling and evaporation	7
Chapter 2 Mechanics	9
2.1 Motion	9
2.2 Momentum and force	10
2.3 Work, energy and power	10
2.4 Friction	11
Chapter 3 Heat and Temperature	13
3.1 Measuring temperature	14
3.2 Thermometric scales	15
3.2.1 The Celsius or Centigrade scale	15
3.2.2 The Fahrenheit scale	15
3.3 Other methods of measuring temperature	15
3.3.1 The air or gas thermometer	15
3.3.2 Using solids to measure temperature	15
3.3.3 Thermocouples	15
3.3.4 Electrical resistance	15
3.3.5 Thermistors	16
3.3.6 Comparison by brightness	16
3.3.7 Infra-red	16
3.4 The Kelvin scale of temperature	16
3.5 Units of heat	17
3.5.1 The Joule	17
3.5.2 The calorie	18
3.5.3 The British thermal unit	18
3.6 Specific Heat	18
3.7 Change of state and latent heat	19
3.7.1 Latent heat of vaporisation	20
3.7.2 Effect of change of pressure on boiling point and latent heat	20
3.7.3 Latent heat of fusion	20
3.7.4 Cooling by evaporation	21

Chapter 4 Thermal Expansion — 23

4.1	Thermal expansion of solids	23
4.1.1	Coefficient of linear expansion	23
4.1.2	Nickel-iron alloy (invar)	24
4.1.3	Allowing for expansion in large metal structures	24
4.1.4	Thermostats	25
4.1.5	Coefficients of superficial and cubical expansion of solids	26
4.2	Thermal expansion of liquids	26
4.2.1	Cubical expansion	26
4.2.2	The effect of expansion on density	26
4.3	The expansion of gases	27
4.3.1	Temperature, pressure, volume	27
4.3.2	The gas laws	27
4.3.2.1	Boyle's Law	27
4.3.2.2	Charles' Law	28
4.3.2.3	The Law of Pressures	28
4.3.2.4	The General Gas Law	29
4.4	The liquefaction of gases	30
4.4.1	Critical temperature and pressure	30
4.4.2	Liquefied gases in cylinders	30
4.5	Sublimation	31

Chapter 5 Heat transmission — 33

5.1	Conduction	33
5.2	Convection	35
5.3	Radiation	36

Chapter 6 The Basis of Chemistry — 39

	The Chemistry of Combustion	39
6.1	The basis of chemistry	39
6.2	Atoms and molecules	39
6.2.1	Compounds and mixtures	40
6.3	Symbols	41
6.3.1	Using symbols to write formulae	41
6.3.2	Radicals	41
6.4	Atomic mass	42
6.5	Molecular mass	42
6.6	Valency	43
6.6.1	Multiple valency	43
6.6.2	Nomenclature	43
6.7	Simple equations	44
6.8	Use of chemical equations	45
6.9	Limitations of chemical equations	46
6.9.1	Reality	46
6.9.2	Physical state	46
6.9.3	Reaction conditions	46
6.9.4	Heat	46
6.9.5	Rate of reaction	46

Chapter 7 Combustion — 47

7.1	The Fire Triangle	47
7.2	Heat of reaction and calorific values	48
7.2.1	Oxidation	48
7.3	What makes a flame a flame?	48
7.4	Laminar flow and turbulent flow	49
7.5	Premixed and Diffusion Flames	50
7.6	Practical examples of premixed flames and diffusion flames	52
7.6.1	The Bunsen Burner	52
7.6.2	A candle flame	52
7.6.3	Flashpoint, firepoint and sustained fires	53
7.6.4	Fireball	53
7.6.5	Vapour cloud explosions	53
7.7	Ignition	54
7.7.1	Spontaneous ignition temperatures	54
7.7.2	Self heating and spontaneous combustion	54
7.7.3	Smouldering	55
7.8	Hazards of oxidising agents	55
7.8.1	Nitric acid and inorganic nitrates	55
7.8.2	Permanganates	56
7.8.3	Chlorates	56
7.8.4	Chromates and dichromates	56
7.8.5	Inorganic peroxides	56
7.8.6	Organic oxidising agents	56
7.8.7	Organic peroxides and hydroperoxides	57

Chapter 8 Simple organic stubstances — 59

8.1	Aliphatic hydrocarbons (paraffins or alkanes)	59
8.2	Unsaturated aliphatic hydrocarbons	60
8.2.1	Olefines or alkenes	60
8.2.2	Acetylenes, or alkynes	61
8.3	Aromatic hydrocarbons	62
8.4	Liquefied petroleum gases (LPG)	63
8.5	Simple oxygen-containing compounds derived from hydrocarbons	64
8.5.1	Alcohols	64
8.5.2	Aldehydes	65
8.5.3	Ketones	65
8.5.4	Carboxylic acids	65
8.5.5	Esters	66
8.5.6	Ethers	66

Chapter 9 Polymers — 69

9.1	Polymers	69
9.2	Fire hazards	70
9.2.1	Toxic and corrosive gases	70
9.2.2	Smoke	71
9.2.3	Burning tars or droplets	71

9.2.4	Exotherms	71
9.2.5	Catalysts	71
9.2.6	Flammable solvents	71
9.2.7	Dusts	72
9.2.8	Self-extinguishing plastics	72
9.3	Monomer hazards	72
	Acrylonitrile, Butadiene, Epichlorhydrin, Methyl Methacrylate, Styrene, Vinyl Acetate, Vinyl Chloride	
9.3.1	Intermediates and hardeners	73
	Isocyanates, Chlorosilanes, Epoxides	

Chapter 10 Other Combustible Solids — 75

10.1	Wood	75
10.2	Coal	75
10.3	Metals	76
10.3.1	Properties of metals	76
10.3.2	Reaction of metals with water or steam	76
10.3.3	Reaction with oxygen	77
10.4	Sulphur	77
10.5	Phosporus	78

Chapter 11 Extiguishing Fires — 79

11.1	Classification of fires by type	79
11.2	Classification of fires by size	80
11.3	Extinguishing fire: Starvation, smothering, cooling	80
11.3.1	Starvation	80
11.3.2	Smothering	82
11.3.3	Cooling	83
11.4	Fire extinguishing media	84
11.4.1	Water	84
11.4.2	Inert gas	84
11.4.3	Foam	84
11.4.4	Vapourising liquids	86
11.4.5	Carbon dioxide and inert gases	86
11.4.6	Dry chemical powders	86
11.4.7	Blanketing	87
11.4.8	Beating out	87

Appendices — 88

A	Metrication: Conversion tables	88
B	Material densities	90

Further reading — 93

Physics and Chemistry
for Firefighters

Physics and Chemistry for Firefighters

Chapter 1

Chapter 1 – Physical properties of matter

Matter is the name given to all material things – anything that has mass and occupies space. Solids, liquids, gases and vapours are all matter. The amount of matter is known as the mass, and is measured in kilograms. In everyday life, the **mass** of a solid is measured in kilograms, although for liquids, gases and vapours, we are more accustomed to using **volume**, the amount of space occupied by a given substance, simply because it is easier to measure. Thus, we talk about litres of petrol and cubic metres of gas. However, gases, vapours and liquids also have mass which can be expressed in kilograms.

Density

Understanding density is extremely important for a firefighter.

Understanding density is extremely important for a firefighter. For example, the density of a gas or vapour determines whether it will tend to rise or sink in air, and be found in the greatest concentrations at the upper or lower levels in a building. The density of a burning liquid partly decides whether it is possible to cover it with water to extinguish the fire, or whether the firefighter will need to use foam or other another extinguishing medium. However, another important factor is how well the burning liquid mixes with water, a property known as miscibility.

Imagine two solid rods, both the same length and width, one made of wood and one from iron. Though they are the same size, the iron rod weighs much more than the wooden rod. The iron rod is said to have a greater density than the wooden one.

The density of a material is defined as the mass of one cubic metre of material. One cubic metre is the standard "unit volume". A unit volume of iron has a greater mass than a unit volume of wood and is thus more dense.

Calculating values for the densities of different substances enables meaningful comparisons to be made. The density of a substance is calculated by dividing the mass of a body by its volume.

$$\text{Density} = \frac{\text{mass}}{\text{volume}}$$

$$\text{In symbols } D = \frac{M}{V}$$

$$\text{so } M = D \times V$$

$$\text{and } V = \frac{M}{D}$$

If mass is measured in kilograms (kg) and the volume in cubic metres (m^3), the "units" of density will be kilograms per cubic metre (kg/m^3). If mass is in grams (g) and volume in cubic centimeters (cm^3), density will be in grams per cubic centimetre (g/cm^3).

(Note: units should always be quoted, and care must be taken not to mix, or confuse, units.)

Water has a density of about 1000 kg/m^3 or 1 g/cm^3.

Physics and Chemistry for Firefighters 1

Mercury has the very high density of 13 600 kg/m³ or 13.6 g/cm³ and is, therefore, 13.6 times as dense as water.

If the density of a substance is lower than the density of water, and does not mix with water, then that substance will float on water. To use our previous example, the density of wood is lower than that of water and the density of iron is higher, so wood floats and iron sinks.

> **If the density of a substance is lower than the density of water, and does not mix with (dissolve in) water, then that substance will float on water.**

The term specific gravity or relative density is sometimes used to give measure of density. The relative density of a substance is the ratio of the mass of any volume of it to the mass of an equal amount of water.

$$\text{Relative density} = \frac{\text{mass of any volume of the substance}}{\text{mass of an equal volume of water}}$$

$$= \frac{\text{density of that substance}}{\text{density of water}}$$

Relative density or specific gravity has no units as the units on the top and bottom of the equation are the same, so they cancel each other out when the one quantity is divided by the other.

1.1 Vapour density

Gases and vapours have very low densities compared with liquids and solids. At normal temperatures and pressures (e.g., 20°C and 1 atmosphere) a cubic metre of water has a mass of about 1000 kg and a cubic metre of air has a mass of around 1.2 kg.

We have previously mentioned that specific gravity is a ratio of the density of the substance in question compared with the density of water, so specific gravity is not a sensible thing to use for gases as their densities are so low: e.g., the specific gravity of air is 0.0013. For this reason, the density of a gas or vapour (vapour density, usually abbreviated to VD) is given in relation to the density of an equal volume of hydrogen, air or oxygen under the same conditions of temperature and pressure.

Hydrogen is often used as a comparison for calculating vapour density because it is the lightest gas. The vapour density of air compared with hydrogen is 14.4, meaning that a given volume of air is 14.4 times heavier than the same volume of hydrogen **at the same temperature and pressure**. For carbon dioxide the vapour density compared with that of hydrogen is 22, so a given volume of carbon dioxide is about 1.5 (that is 22/14.4) times as heavy as the same volume of air at the same temperature and pressure. (If the temperature and pressure are changed, the volume of the gas will change. This will be explained later.)

For fire service purposes it is much more convenient to compare the density of gases and vapours with that of air. The reference gas should always be given to avoid confusion: for example, the vapour density of methane is 0.556 (air = 1), or 8 (hydrogen = 1).

1.2 Liquids of different density

As we have said, the density of a burning liquid partly decides whether it is possible to cover it with water to extinguish the fire, or whether the firefighter will need to use foam or another extinguishing medium.

Consider water poured into two tanks (A) and (B) (Figure 1.1) which are standing on a flat, horizontal surface and are connected by a horizontal pipe as shown. The water will assume equal levels in each tank as, for the system to balance, there must be an equal "head" (or height) of water in each tank above the lowest point of the pipe. The "head" determines the water pressure at any depth: with interconnected tanks as in Figure 1.1, the levels adjust to ensure that the pressures at the level of the pipe are equal.

Figure 1.1 Diagrams showing the difference in density in liquids.

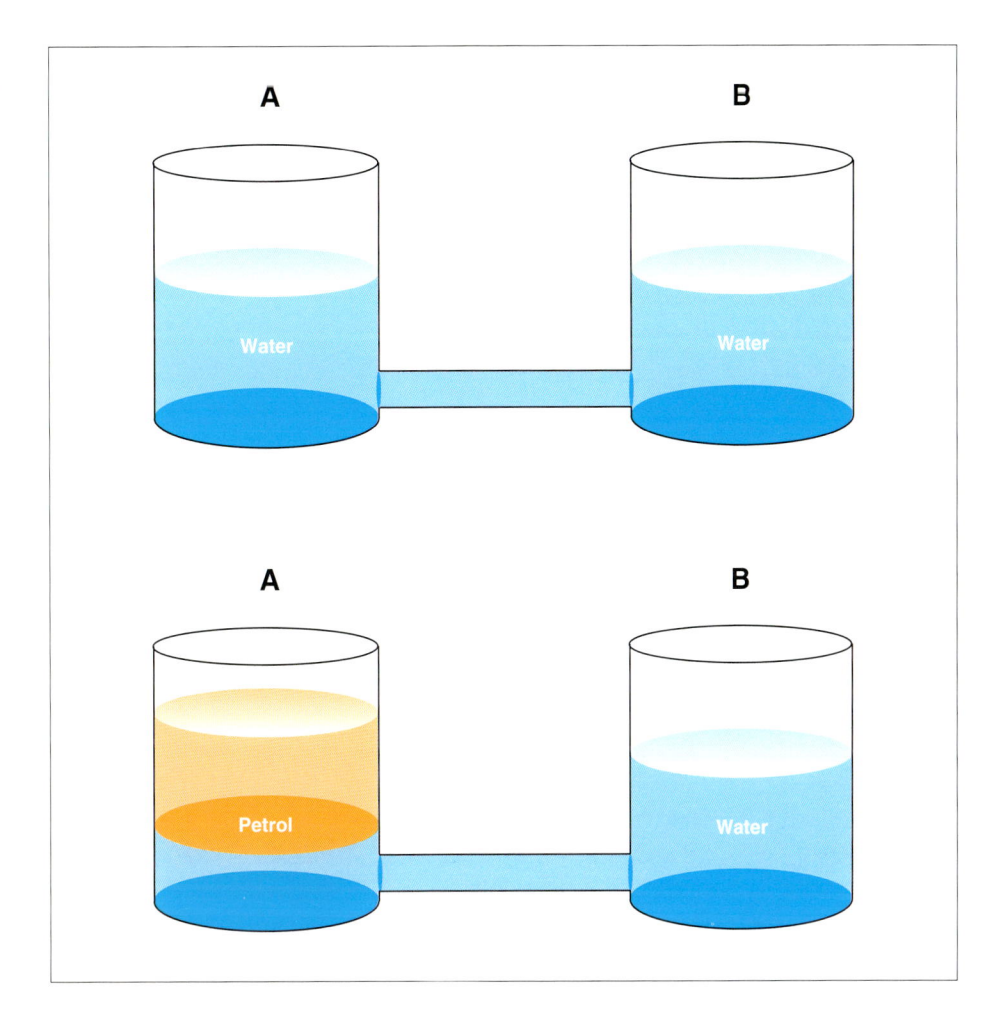

Problems are caused for the firefighter when water is used as an extinguishing agent when burning liquids are present.

As petrol and most other flammable liquids float on water, they cannot be smothered by water, so the addition of water to a fire involving burning liquids may cause the fire to spread further than it otherwise might have done.

Imagine now that petrol is poured into tank (A). As petrol has a specific gravity of about 0.75 and will not mix with water, it will float on the water. For the system to balance, the pressures created by the heads of liquid in each tank must be equal as before, but because petrol is less dense (lighter) than water, a greater "head" of petrol is required to produce the same pressure at depth. Consequently, the level of petrol is higher than the level of water. If water is added to tank (B), the water will raise the petrol in (A) and eventually the petrol will spill over, well before tank (B) is full.

1.3 Gases of different density

Unlike many liquids, all gases and vapours are completely miscible. However, differences in density will affect the way in which they mix. Thus, methane (the main component of natural gas) is a light gas with a vapour density of about 0.5 (air = 1). If it is leaking into a room from a faulty gas appliance, it will rise to the ceiling, entraining (mixing with) air as it rises to form a layer of methane and air mixture which will eventually descend to the level of the leak. (The concentration of methane in the layer will increase as the layer descends.) On the other hand, a leak of propane from a propane cylinder will produce a layer of propane/air mixture at low level in a similar fashion, as the vapour density of propane is roughly 1.5 (air = 1).

All heavier-than air gases, like carbon dioxide (VD 1.53, air = 1) and petrol vapour (VD 2.5, air = 1) will accumulate in low places such as wells and cellars, so creating dangers of asphyxia (suffocation) as well as of fire or explosion in the case of flammable vapours.

Differences in density can also be created by changes in temperature (see Chapter 4). Increase in temperature causes expansion, and a lowering of density. Understanding the consequences of this is extremely important: it can be compared with our example of the behaviour of petrol and water in interconnected tanks. As an example, we can consider the case of a chimney full of the hot products of combustion from an open fire.

The chimney and the rest of the building (Figure 1.2) act rather like our two tanks containing liquids of different densities, in that the chimney is effectively a tank full of hot light gas joined at the base to another tank full of cold, heavy gas, i.e. the surrounding outside air. If we consider the hot, less dense gases inside the chimney and compare them to a column of cold, more dense air outside the chimney we have two volumes of gases of equal height but of different densities, and are thus unbalanced. The hot gases are said to be buoyant with respect to the cold air.

In order to restore the balance, cold air from outside flows into the base of the chimney and drives the hot gas out of the chimney. This would continue until the chimney is full of cold air, but in practice the fire at the base of the chimney continuously replaces the hot chimney gas which is driven from the top of the chimney.

If, however, the flow of hot gas from the chimney is prevented by a cover or damper, a pressure will develop at the top of the chimney, The weight of the column of lighter hot gas in the chimney is not enough to balance the weight of the heavier cold outside air. This unbalanced condition is responsible for a force which drives gas from the top of an open chimney. When the top is closed the force produces a pressure (which can be calculated if the densities of the hot and cold gases are known), and combustion products may escape from the fireplace into the room.

The same thing happens in a burning building. The air inside is heated by the fire and so becomes lighter. It rises and will escape through any available opening provided it can be

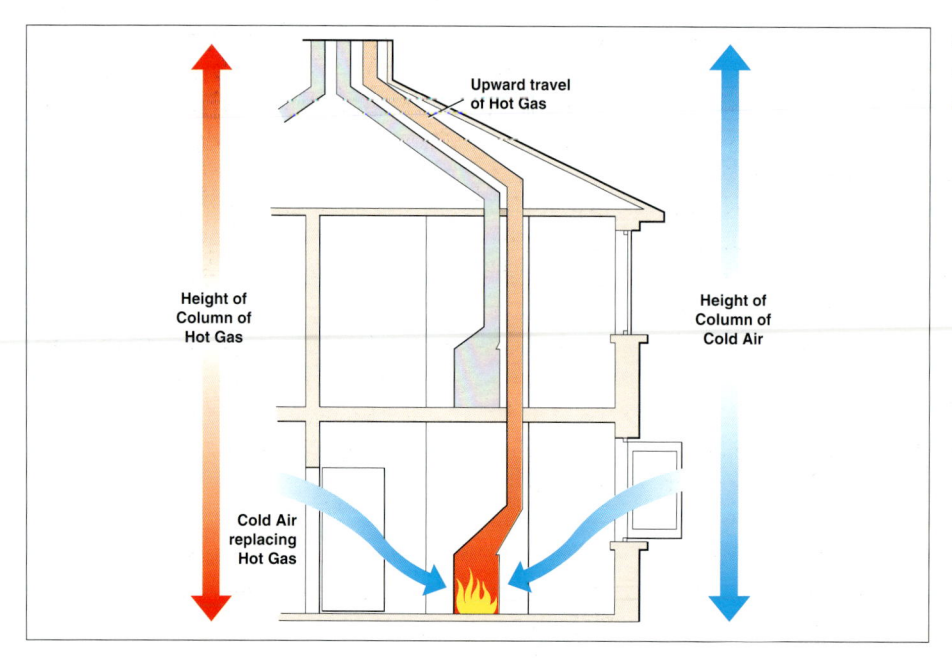

Figure 1.2 Diagram of a chimney showing the travel of convection currents.

replaced by cold air entering at low level (compare with the fireplace). If there is no opening through which it can escape, a pressure will develop in the burning compartment and anyone opening a door or window into this space will release the pressure, which may cause an outrush of hot gases and possibly flames which could engulf them. (See Compartment Fires and Tactical Ventilation. Fire Service Manual Volume 2.)

When there are low and high level openings (e.g., broken windows and a hole in the roof), the building acts as an open chimney, with flames and the hot gases escaping at high level with cold fresh air entering from below. Under these circumstances, the fire will be very intense.

1.4 Matter and energy

Matter can exist in three states: **solid**, **liquid** or **gas**. Some substances are quite commonly found in all three states – for example, water is found as ice, liquid water and water vapour (steam) – but most substances are, at normal temperatures, found only in one or two of the states. For example, steel is solid up to its melting point of around 1400°C (the melting point varies according to the composition of the steel). Its boiling point, the point at which it turns into a vapour, is about 3000°C. Carbon dioxide is normally a gas, but under pressure it can be liquefied and if it is cooled sufficiently, it solidifies. Oxygen is normally a gas but it can be liquefied at very low temperatures (boiling point -183°C).

All matter is made up of extremely small particles called **atoms**. Atoms have a central core, or **nucleus** which contains smaller particles called **protons** and **neutrons**. Protons possess a positive electrical charge. The nucleus is surrounded by a system of **electrons**, which each carry a negative electrical charge. Atoms contain as many electrons as protons. As the number of protons and electrons are matched, and each proton possesses an equal and opposite charge to each electron, atoms are electrically neutral. The number and arrangement of electrons around the nucleus determines the chemical behaviour of the atom, that is to say, it determines which other atoms it will combine with. Chemical reactions take place when electrons move between atoms. An atom which has lost or gained one or more electrons in a chemical interaction will possess an electrical charge and is called either a positive or negative **ion**.

It now seems that it is not possible to get more than about 92 protons in a nucleus without it becoming so unstable that it falls apart. Otherwise, any number of protons is possible.

> **An element is a substance which contains atoms which are all of the same type.**

An element is a substance which contains atoms which are all of the same type: they all have the same number of protons. As there can be anything up to just over 92 protons in an atom, there are, just over 92 stable elements. Atoms of different elements can combine to form **molecules**.

Some molecules consist of two or more atoms of the same kind: for example, an oxygen molecule consists of two oxygen atoms (O_2). Other molecules consist of two or more atoms of different kinds: carbon dioxide consists of two atoms of oxygen and one of carbon (CO_2), water consists of two atoms of hydrogen and one of oxygen (H_2O). Carbon dioxide and water are chemical **compounds** and they can, by chemical means, be split into their component elements. This forms the basis of the science of chemistry, which is dealt with in Chapter 6.

Atom is a Greek word meaning 'indivisible': until about 70 years ago, it was believed that atoms could not be split into smaller particles. We know now that this is wrong and that atoms of one element can be "split" or combined with other particles to make new atoms of other elements. (The subject of atomic physics is discussed in the Manual of Firemanship, Part 6c, Chapter 45, Section 11.)

Energy is expended in doing work and may be in one of a number of different forms. **Heat**, **light** and **electrical energy** are well known from everyday

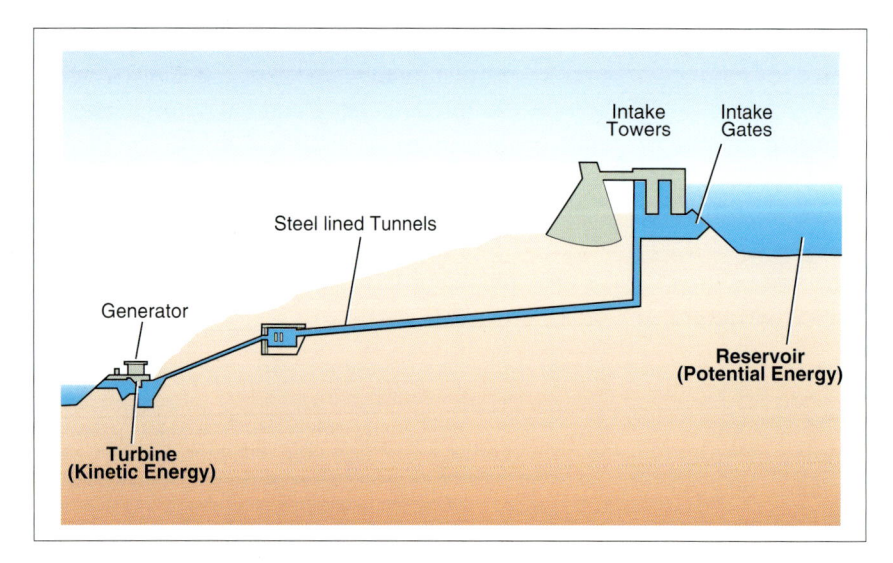

Figure 1.3 Diagram showing the conversion of potential into kinetic energy.
(Diagram: North Hydro)

experience. There is also **potential energy**, which is possessed by a body due to its position, for example by water stored in a hydroelectric dam, and **kinetic energy**, which is energy possessed by a moving body, for example by the water from the dam flowing through the turbines in the turbine hall). The potential energy is converted into kinetic energy as the water flows under gravity to the turbine hall, where it is then converted into electrical energy (Figure 1.3).

> **The faster something is moving, the more kinetic energy it has.**

For the vast majority of purposes we can say that energy cannot be created or destroyed – it can only be converted into another form of energy. (Note that when some radioactive processes occur, minute quantities of mass are "lost' and converted into large amounts of energy.)

The firefighter is mostly concerned with energy in the form of heat. Heat may be produced by a chemical change, such as combustion, in which we say that **chemical energy** is released as heat energy. Mechanical energy or kinetic energy can also be converted into heat energy by friction (e.g., frictional heating of brake pads).

We are familiar with the concept of temperature from everyday experience. **Temperature** is a measure of how hot something is, and is related to how "fast" the constituent molecules are moving.

> **Temperature also determines which way heat will flow. Heat can only move from something at a high temperature to something at a lower temperature.**

The molecules which make up any substance, even a solid, are continually moving, although in a solid they vibrate around a fixed position. They also exert a force of attraction to each other, which becomes greater the closer they are together. The movement of molecules tends to spread them out while the attractive force, or force of cohesion, tends to bind them together.

If a solid is heated, heat energy is stored in the substance as the vibrational energy of the molecules. As more energy is stored, they vibrate faster and take up more space. At the same time, the temperature of the solid rises and thermal expansion occurs.

A temperature is reached when the molecules are vibrating so much that they break free of the rigid framework in which they have been held by the cohesive forces and become free enough to slide past each other, although they do not have complete freedom. At this point the solid melts and becomes a liquid.

Further heating causes the temperature to increase, and the energy is stored as kinetic energy of the molecules, which move with increasing rapidity until they are moving fast enough to overcome the cohesive forces completely. At this point, the liquid boils and turns into a gas (or more correctly, a vapour). If heat is taken away from a substance, kinetic energy of the molecules decreases and the reverse processes occur.

1.5 Melting, boiling and evaporation

The temperature at which a solid turns into a liquid is called the **melting point**. If we are considering a liquid turning into a solid, the temperature is called the **freezing point**, though these two temperatures are the same for the same substance under the same conditions. The temperature at which a liquid boils and becomes a vapour is the **boiling point**.

Since energy is required to overcome the forces of cohesion when a substance melts or boils, the heat which is supplied during these processes does not cause a rise in temperature of the substance. Conversely, when a vapour condenses or a liquid solidifies, it gives up heat without any fall in temperature so long as the change is taking place. So, melting or freezing, for a given substance at a given atmospheric pressure, take place at a certain temperature: for the transition between water and ice, at normal atmospheric pressure, this takes place at 0°C. Boiling, for a given substance at a given atmospheric pressure take place at another certain temperature: for the transition between water and steam, at normal atmospheric pressure, this takes place at 100°C.

Even at temperatures below boiling point, some molecules at the surface of the liquid may gain enough energy from colliding with other molecules for them to escape into the surrounding space as vapour. This process is **evaporation**.

Imagine a liquid in an enclosed space, where there is already air, such as water in a saucepan. Even if the pan is not heated, evaporation will take place.

The evaporating molecules build up a pressure known as the vapour pressure. At the same time, some molecules will re-enter the liquid. For any given temperature below the boiling point there is a definite vapour pressure at which the number of molecules which escape is just balanced by the number which are recaptured by the liquid.

Boiling occurs when the vapour pressure has become equal to the surrounding atmospheric pressure, the pressure of air. Vapour then forms not only at the surface of the liquid, but also in the body of the liquid, and we see bubbles.

If the external pressure is increased, the vapour pressure at which boiling will take place is increased and so the temperature must increase. If the external pressure falls, the reverse is true and the temperature at which boiling occurs will be lower.

Physics and Chemistry for Firefighters

Chapter 2

Chapter 2 – Mechanics

This Chapter will discuss how things move and how the movement of objects is linked to their mass, the energy they possess and the forces which act on them.

2.1 Motion

Imagine a body moving from a starting point A to another point B: for example a car moving between two cities.

The distance travelled by the car is the length of the line X. The average speed the car travelled at will be the distance travelled by the car divided by the time taken.

$$\text{Speed} = \frac{\text{distance travelled}}{\text{time taken}}$$

It has units of metres per second (m/s), or if distance is measured in miles, and time in hours, the units will be miles/hour (mph).

Although people tend to use speed and velocity interchangeably, there is a difference between them. Velocity has a direction associated with it: it is what we call a **vector** quantity.

The line Y shows the length of a straight line connecting the starting and finishing point. The length and direction of this line together give the **displacement** of the car from the starting point. Displacement is also a vector quantity.

Imagine a car journey from London to Bristol. Bristol is 200 km west of London, so at the end of the journey the car has a *displacement* of 200 km from London. However, the *distance* travelled will be longer than this though, as the roads tend to take convenient routes through the countryside rather than following perfectly straight lines. There may even be times when the car is travelling North-South rather than East West as it follows the road. At those times, the displacement of the car from London is not increasing, so although it has speed, its velocity is zero.

Figure 2.1 Car moving from A to B.

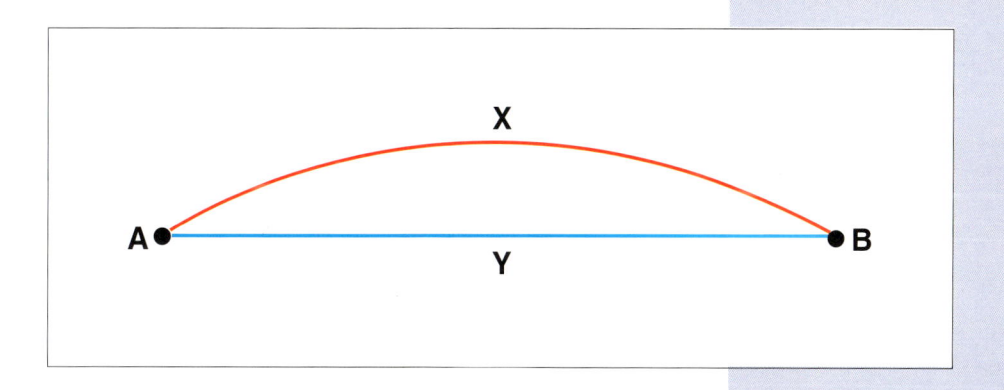

Physics and Chemistry for Firefighters 9

> **Speed is the rate of change of distance.**
>
> **Velocity is the rate of change of displacement.**
>
> **Acceleration is the rate of change of velocity.**

The units of speed and velocity are m/s, while those for acceleration are m/s^2.

2.2　Momentum and force

Momentum is the product of mass and velocity

$$\text{Momentum} = \text{mass} \times \text{velocity}$$

So, a 2 kg object travelling at 10 m/s has a momentum of 20 kg.m/s

Force is the product of mass and acceleration

$$\text{Force} = \text{mass} \times \text{acceleration}$$
$$F = m\,a$$

The units are kg.m/s^2, which known as the **Newton (N)**.

Imagine that our 2 kg object is initially at rest, and then a force is applied which makes it accelerate at 10 m/s^2. The applied force is equal to the body's mass times its acceleration.

$$F = ma$$
$$F = 2 \times 10 \text{ kg.m/s}^2$$
$$F = 20 \text{ N}$$

The "force of gravity" provides an acceleration which acts on everything. At the earth's surface, anything which is dropped will accelerate, under gravity, at 9.81 m/s^2 (this quantity is often referred to by the symbol g). It also determines the forces that are responsible for the movement of hot, buoyant gases in fires, and as described in Chapter 1.

The weight of a body is a measure of how strong the force due to gravity is on an object. In everyday speech, we usually use mass and weight interchangeably, but just as the words "speed" and "velocity" have different meanings, weight and mass also have different meanings.

Weight is the force due to gravity which acts on an object. The acceleration due to gravity that an object will experience is $g = 9.81$ m/s, so that a mass of two kilograms experiences a force due to gravity of

$$F = m\,a$$

In this case $F = 2 \times 9.81$
i.e. the weight of the object is 19.62 N

Giving definitions of weight and velocity which are different from those used in every day speech may seem like an unnecessary complication, but much science and engineering is only possible if every aspect of a problem is precisely defined, and every scientist or engineer everywhere knows exactly what is meant by each word.

2.3　Work, energy and power

Work is another every day term which has a rigid definition in science. If a body moves because a force acts on it, we say that work is being done on that object. Work is only being done if the object moves – if a force is being applied just to keep an object stationary, no work is done. The mathematical expression of work is:

$$W = F\,s$$

where W is the work done (in Joules), F is the constant force applied (Newtons) and s is the distance moved in the direction of the force (in metres). (The Joule (J) has units N.m, or $(kg.m/s^2)$.m, i.e., $kg.m^2/s^2$: this illustrates the importance of ensuring that a consistent set of units is used.)

Taking our 2 kg mass, if this is raised vertically through 2 m, the work done *against gravity* will be

$$W = F s$$
$$W = 2 \times 9.81 \times 2$$
$$= 39.24 \text{ J}$$

Note that if the force acts in a different direction to the direction of travel, less work will be done. The interested reader should refer to the various text books on this topic.

If something is capable of doing work, it possesses energy. The energy that a body possesses by virtue of the fact that it is moving is called kinetic energy. The kinetic energy that a body possesses is equal to the amount of work that must have been done to it to increase its velocity from zero to whatever velocity it has. For a mass m moving with velocity v:

$$\text{Kinetic energy} = \frac{1}{2} mv^2$$

The units are $kg.m^2/s^2$, i.e., Joules.

Potential energy is the amount of energy that an object possesses because of its position or the arrangement of its components.

A body held above the ground has potential energy because it could do work while it is falling. Our 2 kg mass raised vertically through 2 m has a potential energy equal to the work required to raise it to that height, i.e., mass × acceleration × height (abbreviated as mgh), which as we have seen has the units of Joules. This is a consequence of the position of the mass in the earth's gravitational field. Other forms of energy are encountered, e.g., an electron in an electric field has electrical potential energy, while a stretched rubber band has elastic potential energy.

2.4 Friction

When an object moves or tries to move over a surface, both the object and the surface experience a frictional force along the common surface, each in a direction which opposes the relative motion of the surfaces.

Even on two very flat, well-polished surfaces, there are many microscopic imperfections which make the contact area much smaller than it would seem. These imperfections may interlock and, in the case of metal, may even weld together under very high local pressures.

We can see, then, that this locking and bonding will inhibit motion and that energy will be needed to overcome the frictional force. This energy appears as heat: when energy is expended on overcoming friction, heating occurs. This is the principle behind the rough strip on match boxes which provides the heat to start the chemical reaction in a match, and is also why brake blocks get hot on a car or bicycle.

Physics and Chemistry for Firefighters

Chapter 3

Chapter 3 – Heat and Temperature

Some of the most destructive effects of fire are caused by heat, so it is obvious that the fire fighter should understand the effect of heat on materials. This chapter will discuss heat and temperature, how the two are linked and will lay the foundations for a discussion of the reaction of solids, liquids and gases to changes in temperature.

**The amount of heat energy in a body cannot be measured directly. When heat energy is supplied to a body, it is true that the body's temperature will rise.
HOWEVER, temperature is just a measure of how hot a body is, NOT the amount of heat energy it contains.**

In Chapter 1 and 2 it was stated that:

Energy

- is the ability to do work;
- can neither be created nor destroyed; and
- can exist in a number of different forms.

Heat is one form of energy. It can be produced by chemical means, for example by burning coal or oil, or by mechanical means, by friction. Passing a current through an electrical resistance also produces heat (e.g., an electric fire).

Heat can be converted into other forms of energy, for example into pressure energy in a steam boiler.

It is also possible to convert heat back into chemical energy or electrical energy.

Heat always flows from high temperature to low temperature. If a hot body and a cold body are placed in contact, the hot body (the one at the higher temperature) loses heat and the cold body (the one at the lower temperature) gains heat.

The fact that heat and temperature are not the same thing can be seen from a simple experiment. Imagine a piece of fine copper wire held in the flame of a match. After a couple of seconds, the wire will glow red hot, which tells us that the temperature of the wire has increased to 800 – 900°C.

Now imagine a similar match held under a kettle containing a litre of water. There would be no noticeable change in the temperature of the water, yet the amount of heat supplied to the wire and supplied to the water would be roughly the same.

The rise in temperature in a body to which heat is supplied is decided by three factors: the amount of heat supplied (or "transferred") to the body, the mass of the body and the *specific heat capacity* of the material from which the body is made.

The meaning of the term 'specific heat capacity' will be discussed later.

3.1 Measuring temperature

> **The human body cannot measure temperature, it can only make comparisons.**

The human body cannot tell reliably whether something is hot or cold, it can only compare what it is currently feeling with what it felt immediately beforehand. If you place one hand in a bowl of cold water and the other in a bowl of hot water and then, after an interval, both hands are placed in a bowl of tepid water, the hand which was in the bowl of cold water will feel that the tepid water is 'hot', while that from the hot water will feel that the tepid water is 'cold'.

Because it can only make comparisons, the human body cannot give a numerical value to temperature – people cannot step out into the street and reliably measure the air temperature just by the feel of the air on their skin.

Temperature can, though, be measured by making use of one of the effects of heat on materials. The commonest example is the use of the way that liquids expand as their temperature rises, the property of thermal expansion of a liquid. This is the principle behind the **thermometer** (Figure 3.1).

The thermometer consists of a narrow tube of fine bore with a small bulb at one end, and sealed at the other, containing a suitable liquid. The liquid is most commonly mercury, which has the advantages of a high boiling point (357°C), a uniform expansion coefficient and a low heat capacity; it is also opaque. However, its freezing point is about -39°C which makes it unsuitable for measuring temperatures greatly below the freezing point of water. Alcohol has a lower freezing point (-112°C) but also a much lower boiling point (78°C) than mercury. It can be is used for low temperatures work. Coloured water is sometimes used for rough measurement of temperatures between the freezing point and boiling point of water.

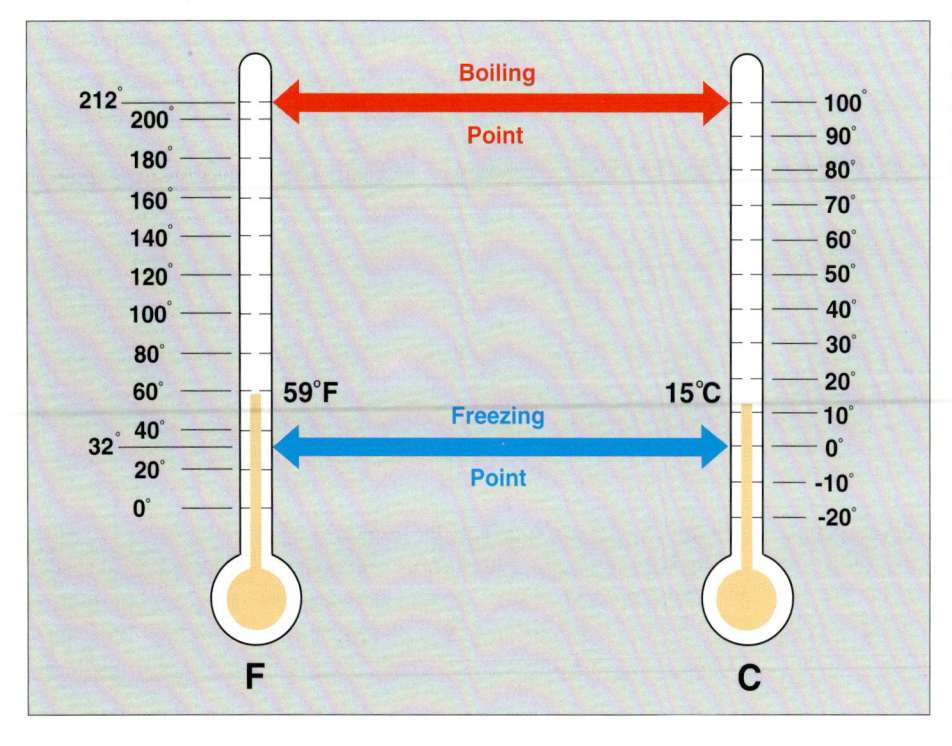

Figure 3.1 Diagram showing thermometer and a comparison between the Celsius and Fahrenheit thermometer scales.

3.2 Thermometric scales

> **Two scales are in common use for measuring temperature: the Celsius scale and the Fahrenheit Scale.**

Two fixed points are required for the construction of a thermometric scale. For the Celsius (or Centrigrade) scale of temperature, the melting point of pure ice and the boiling point of pure water are taken as the fixed points *at standard atmospheric pressure*.

So, to fix the lower point of the scale, the bulb of the thermometer is placed in melting ice, while the upper fixed point of the thermometer is determined by placing the bulb in the steam above the surface of boiling water (at standard atmospheric pressure). If the pressure is different from the standard atmospheric pressure, a correction has to be applied to the upper fixed point. The level at which the liquid in the thermometer stands at each of the fixed points is marked on the stem of the thermometer.

Two thermometric scales are in common use:

3.2.1 The Celsius (or Centigrade) scale

On this scale the lower fixed point is marked 0. The upper fixed point is marked 100. The stem between these two points is divided into 100 equal divisions or degrees. These divisions are called Celsius degrees.

3.2.2 The Fahrenheit scale

The inventor of this scale used a freezing mixture to give him his lower fixed point, and the boiling point of water for the upper fixed point. The scale was divided into 212 equal divisions, which gave the freezing point of water as 32°F. There are 180 Fahrenheit degrees between the freezing point and the boiling point of water (at standard atmospheric pressure).

3.3 Other methods of measuring temperature

The 'liquid-in-glass' thermometer is not the only method of measuring temperature. There are several other methods, including the following:

3.3.1 The air or gas thermometer

Instead of using a liquid, a bulb containing air or some other gas can be used. In one such thermometer, the expansion of the gas causes a short thread of mercury to move along a scale. These thermometers are very sensitive, but may require correction to compensate for atmospheric pressure.

3.3.2 Using solids to measure temperature

The way that a solid expands when its temperature rises can be used for temperature measurement. The expansion may be used directly, or the differing expansion of two dissimilar metals may be used. This will be discussed further in the next Chapter.

3.3.3 Thermocouples

When the junction of wires of two different metals (for example iron and copper) is heated, an electrical potential (a voltage) appears at the junction. A calibration can be made between the potential and temperature, so that temperature can be measured indirectly by measuring the potential with a sensitive voltmeter. There are various types of thermocouple, some of which are capable of recording extremely high temperatures. (See Figure 3.2)

Thermocouples junctions can be made very small and so only take a short time to heat up or cool down. They are very good for following very rapid changes in temperature.

3.3.4 Electrical resistance

The electrical resistance of a wire increases with a rise in temperature and the change of resistance may be used to measure temperature. Platinum is normally used as it has a high melting point and a high 'temperature coefficient of resistance', so that a small rise in temperature produces a (relatively)

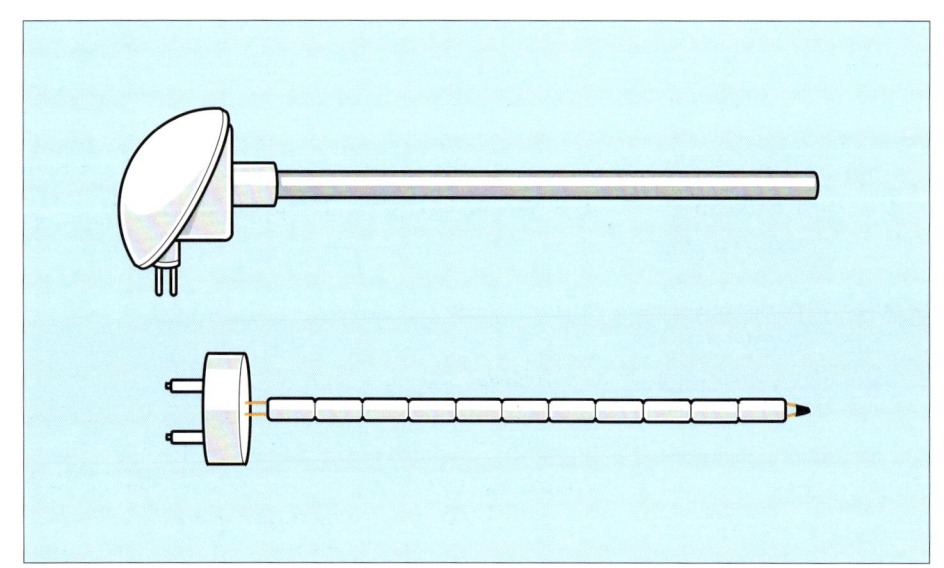

Figure 3.2 Sketch of a pyrometer of the thermo-couple type. The heat sensitive element (below) is housed in a container (above) which protects it from the effects of heat and mechanical wear and tear.

large rise in resistance. Platinum resistance thermometers can measure between -200°C and 1200°C. Their disadvantage is that they are large compared with thermocouples and so do not follow rapid changes in temperature very easily.

3.3.5 Thermistors

Thermistors are semiconductor devices, which have a negative temperature coefficient of resistance, so an increase in temperature produces a decrease in resistance. They are very robust and can be made very small and so can follow rapid changes in temperature. Their range is generally from -70°C to 300°C, but they are less accurate than resistance thermometers.

3.3.6 Comparison by brightness

At temperatures above about 750°C, objects start to glow, first a dull red, changing gradually to yellow and brightening as the temperature is raised to about 1250°C. Temperature measurements can be made by comparing the brightness of the object with the filament of an electric lamp whose brightness can be altered by varying the current flowing through it. If the current is too low, the filament appears darker than the object, while if it is too large, the filament appears brighter. When the filament "disappears" against the object, they have the same brightness *and* temperature. The latter can be found indirectly by measuring the current through the filament.

3.3.7 Infra-red

Infra-red cameras and other sensors detect heat in the same way that our eyes detect light. Light represents only one portion of the *electromagnetic spectrum* – a rainbow of different types of radiation which includes, in addition to visible light, radio waves, microwaves, ultra-violet radiation and X-rays.

Infra-red radiation is given off by bodies (and some gases) when they are hot. Infra-red sensors are sensitive to this type of radiation and can be designed to measure the temperature of an object by analysing the strength and the wavelength of the radiation.

Infra-red radiation behaves in exactly the same way as light, but it can pass through some things that light can't and it is blocked by some things that light can pass though. In particular, infra-red can pass through smoke at concentrations which block visible light. This is why infra-red cameras have become so valuable in search and rescue operations. The infra-red radiation emitted by an unconscious person lying in a cool environment can easily be detected.

3.4 The Kelvin scale of temperature

The Kelvin or 'Absolute' scale of temperature starts at -273°C, which has been found (theoretically and experimentally) to be the lowest temperature that it is possible to achieve. We have discussed before in the sections on melting and boiling, that the hotter a mass is, the faster the molecules that make up that body are moving. At -273°C, the molecules that make up a substance stop moving and it is not possible to cool the mass any further.

The Kelvin or absolute scale of temperature has its zero at -273°C. Degrees on this scale are the same size as Celsius degrees, and are denoted by the symbol K, so

> Absolute temp. = Celsius temp. + 273
> 0 K = -273°C
> 273 K = 0°C
> 100°C = 373 K

Although the Celsius scale is the most widely used, the Kelvin scale must be used in certain circumstances – particularly when calculating how the volume of a gas changes with temperature and pressure (see Chapter 4). This will be discussed in more detail later, but Figure 3.3 shows how a given mass of gas will occupy a smaller volume as it cools (assuming that it does not condense into a liquid). For most purposes, it is possible to assume that the volume of this mass of gas is proportional to its temperature in *degrees Kelvin*. Therefore, if the temperature is doubled at a given pressure, its volume will double. Conversely, if the temperature is halved, the volume will halve. In principle, the volume would become zero at 0 K if it remained as a gas. However, all real gases will condense to liquid or solid form at low temperatures.

Note that in equations, the symbol T is normally used for temperature, but great care must be taken in remembering which temperature scale is being used.

3.5 Units of heat

In the same way that length is measured in metres and temperature is measured in degrees, there are units which are used to measure the amount of energy in a body – what is colloquially called "heat". The concept of energy was introduced in Chapter 2, and discussed in the context of "work done". Energy can neither be created or destroyed, but can be converted from one form into another, e.g., potential energy into kinetic energy, or chemical energy into electrical energy. The conversion process is never 100% efficient, and some of the energy will appear in a different form, most commonly as "heat". Thus, some of the energy expended in bringing a moving car to rest is dissipated as heat generated by friction at the points of contact between the brake shoes and the brake drum.

3.5.1 The Joule (J)

The unit that scientists and engineers use to measure heat is the same as that used for energy, i.e., the **Joule**, which is named after a nineteenth century Manchester brewer who became interested in how much energy was needed to heat water. It is defined from mechanics; where energy is the abil-

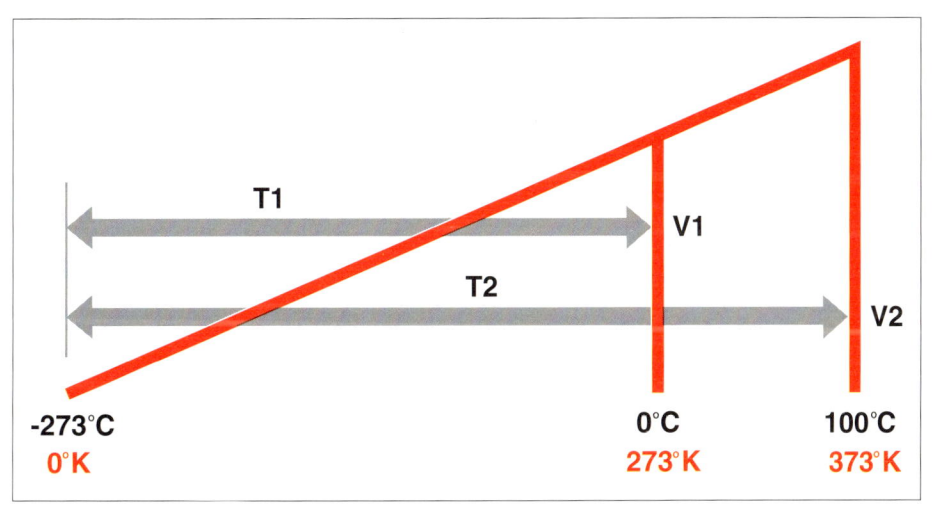

Figure 3.3 Graph showing the zero on the Kelvin or absolute scale.

ity to perform work, so work and energy are measured in the same units, as in Chapter 2. One Joule of work is done when the point at which a 1 Newton (1 N) force is applied moves through 1 metre in the direction of the force.

For convenience, to save writing strings of zeroes, larger units based on the Joule are used, the kilojoule and the megajoule:

> 1 kilojoule (1 kJ) = 1 000 J
> 1 megajoule (1 MJ) = 1 000 kJ = 1 000 000 J

In the same way that feet and inches were replaced by metres, there are older units which have now been replaced by the Joule. These will now be discussed briefly.

3.5.2 The calorie

This is defined as the quantity of heat required to raise the temperature of 1 gram of water through 1°C. The energy content of food is often measured in calories, though, confusingly the number of 'calories' we talk about in a bar of chocolate in everyday speech is actually the number of kilocalories.

> 1 kilocalorie = 1000 calories.
> 1 calorie = 4.18 joules.

3.5.3 The British thermal unit (Btu)

This is the quantity of heat required to raise the temperature of 1 lb of water through 1°F.

Another British unit is the **therm**, which is equal to 100 000 Btu (10^5 Btu).

> 1 Btu = 1 055 J
> 1 therm = 105 500 kJ.

3.6 Specific heat

As we have discussed, heat energy can only flow from a body at a higher temperature to one at a lower temperature. Heat transfer continues either until both bodies are at the same temperature (i.e., the initially hot body has cooled and the initially cold body has warmed), or until the two bodies are separated. Whatever, a certain amount of heat energy will have been transferred, which can be measured in Joules.

When heat is added to a body the temperature rises. The rise in temperature of the body depends on three things:

- the amount of heat energy supplied to the body;
- the mass of the body; and
- the specific heat capacity of the body.

The **specific heat capacity** of a material is the heat required to raise the temperature of one kilogram of the material by 1°C, and so is measured in Joules per kilogram per degree centigrade (J/kg°C). **In equations, the letter c is used as a symbol for specific heat capacity**.

Some texts may discuss the **heat capacity** of an object. This is the specific heat capacity of the substance it is made from multiplied by the mass. They are different quantities though easily confused. In general, the 'specific' is used when the value under discussion refers to a unit mass of material.

Imagine two containers, one containing water and the other containing the same mass of oil. Now imagine that a given amount of heat energy is supplied to each. This could be done by placing the containers on identical burners – ones that supply heat at the same rate – for the same amount of time.

Both containers are equipped with thermometers, so that the temperature rise can be observed.

Now, after 3 minutes, it is found that the temperature of the oil has risen by 10°C, but the temperature of the water has only risen by 5°C in the same time.

The masses of the oil and water are the same, and the amount of heat energy supplied to the oil and water was the same, so the difference in the temperature rise experienced by the two samples must be due to a difference in specific heat capacity.

The same amount of heat caused a greater rise in temperature in the oil than in the water. We know that specific heat capacity is a measure of how much heat it takes to bring about a one degree rise in temperature in a given material. Therefore, the water has a greater specific heat capacity than the oil as the water showed a smaller temperature rise for a given amount of heat supplied.

> **In summary:**
> **The larger the specific heat capacity of a substance, the more energy it takes to raise the temperature by a given amount.**

Table 3.1: *Specific heat capacities of some materials*

Material	c (J/kg°C)
Water	4200
Iron	460
Aluminium	900
Copper	400
Mercury	140
Glass (ordinary)	670
Ice	2100
Earth, rock, etc.	840
Carbon tetrachloride	850
Methylated spirit	2400
Benzene	1720
Glycerol	2560

The other side to this statement is that materials with a low specific heat capacity will heat up more rapidly in a fire situation than those of high specific heat capacity.

Water has an unusually high specific heat capacity: 4 200 J/kg per °C. There are very few substances which have a higher value than this, the most notable being hydrogen at constant volume, and mixtures of certain alcohols with water.

> This is one of the reasons why water is good for fighting fire – a given mass of water can absorb a relatively large amount of heat energy.

Some values of specific heat capacities are shown in Table 3.1.

Substances such as petrol, alcohol and the like have low specific heat capacities. They are also readily vaporised and may produce hazardous vapours. In general, combustible materials with low specific heat capacities are of capable of promoting fire risks.

[It should also be noted that the *surfaces* of solids of low density, such as polyurethane foam, also heat up very rapidly when exposed to a heat transfer process (e.g., radiant heating from an electric fire). This is a result of low *thermal conductivity*. Instead of heat being rapidly transferred into the body of the solid by conduction, it "accumulates" at the surface, resulting in a rapid temperature rise. As a consequence low density combustible materials can be ignited very much more easily than materials of high density.]

3.7 Change of state and latent heat

By 'change of state' we mean the changes between the solid and liquid states (melting/freezing), liquid and gas states (boiling/condensation), and for a relatively few pure compounds, solid and gas states (sublimation). Freezing, melting, boiling, condensation and sublimation all cause a change of state.

3.7.1 Latent heat of vaporisation

When a kettle is put on to boil, heat enters the water from the kettle element and the temperature of the water rises until it reaches 100°C.

At this temperature the water boils, that is to say bubbles of vapour form at the bottom and rise to the surface where they burst and escape as steam. Once the water has started to boil, the temperature *remains constant* at 100°C – it doesn't get any hotter than 100°C. However, heat energy continues flow from the element into the water. This energy is not increasing the temperature of the water, but is being used to allow the water molecules to pull themselves apart from each other, converting the water from the liquid state to the vapour state, i.e., from liquid water into water vapour (steam).

Experiments show that 2 260 000 Joules (2.26MJ) are required to convert 1 kilogram of water at its boiling point into steam at the same temperature. This is known as **the specific latent heat of vaporisation** for water (latent means hidden – the latent heat is hidden heat because it doesn't cause a temperature rise). This extra heat goes into the vapour, but does not indicate its presence by producing a rise in temperature.

> It is this large amount of latent heat which makes water mists and sprays so effective in extinguishing flames – vaporisation of the water droplets helps cool the flame, and cause it to extinguish.

When steam condenses to form liquid water, the same amount of latent heat is given out. This is why steam can cause serious burns.

> **The specific latent heat of vaporisation of a substance is the amount of heat energy needed to change a unit mass of the substance from the liquid to vapour without a temperature change.**

All liquids besides water absorb latent heat when they are turned into vapour. For example, 860 000 J are required to convert 1kg of alcohol into vapour at its boiling point.

Latent heat is measured in Joules per kilogram (J/kg), although it is more usually expressed in kilojoules per kilogram (kJ/kg).

3.7.2 Effect of change of pressure on boiling point and latent heat

Water 'normally' boils at 100°C. By 'normally' we mean that it boils at 100°C when the external air pressure is the standard atmospheric pressure of 1.013 bars where 1 bar = 10^5 N/m^2. This is the pressure of air which will support 760 mm of mercury in a mercury barometer, and so is sometimes written as "760mm mercury" or "760mm Hg".

> **If the external pressure is raised, the boiling point is raised, and if the external pressure is lowered, the boiling point is lowered.**

This effect is used in pressure cookers and in pressurised cooling systems for car engines, where the increased pressure raises the boiling point of the liquid in the system.

This behaviour is also used in the storage of "liquefied gases" such as propane and butane. At increased pressures, these gases liquefy at normal temperatures, and allow large amounts of "gas" to be stored in a relatively small volume (see Section 4.4).

Raising the boiling point increases the quantity of heat needed to raise the temperature of the cold liquid to the new boiling point, but it decreases the latent heat of vaporisation.

3.7.3 Latent heat of fusion

Just as latent heat is taken in when water changes to vapour at the same temperature, a similar thing happens when ice melts to form water. In this case, the latent heat is not so great. It requires 336 000 J (336 kJ) to convert 1 kg of ice at 0°C to water at the same temperature. Likewise, when water at 0°C freezes into ice, the same quantity of heat is given out for every 1 kg of ice formed. This is called the specific latent heat of fusion of ice. This is not confined to water alone; other substances absorb latent heat when they melt and conversely they give out latent heat on solidifying. This is the latent heat of fusion.

The definition of the specific latent heat of fusion of a substance is the quantity of heat required to convert unit mass of the substance from the solid to the liquid state without change in temperature. The same units (J/kg or kJ/kg, etc.) are used as for the latent heat of vaporisation.

3.7.4 Cooling by evaporation

Some liquids have a low boiling point and thus change from liquid to vapour quite easily at ordinary temperatures: these are called volatile liquids. Methylated spirit and ether are of this type. If you drop a little methylated spirit or ether onto your hand, it evaporates rapidly and your hand feels cold. Some local anaesthetics work in this way, 'freezing' the pain.

The cooling is brought about because, to change from liquid to vapour, the liquid absorbs heat energy from the hand to provide the latent heat of vaporisation of the liquid. The hand therefore feels cold. Water would also cause the hand to become cold but not so noticeably as methylated spirit. The spirit has a lower boiling point than water and so it evaporates more quickly at the temperature of the hand.

Physics and Chemistry for Firefighters

Chapter 4

Chapter 4 – Thermal Expansion

> **In this Chapter we will discuss the practical problems and uses of thermal expansion and the ways in which we can calculate the degree of thermal expansion that a material will experience.**

4.1 The thermal expansion of solids

As we have discussed in Chapter One and Chapter Three, a substance, whether solid, liquid or gas, will tend to expand when it is heated as long as it is not constrained by a container, a change of state or a change in chemical composition.

When a solid is heated, it expands in all three dimensions and, therefore, increases in length, breadth and thickness. The increase in length is often the most important, although the increase in area and volume due to thermal expansion can be readily calculated by considering the increase in each dimension

Within normal ranges of temperature, a solid which is "homogeneous" in structure, such as an iron bar, expands uniformly: the expansion of a bar in each direction is proportional to the rise in temperature.

> **"homogeneous": properties are the same in all directions**

The expansion is also proportional to the length of the bar, but varies with the nature of the substance of which the bar is made.

4.1.1 Coefficient of linear expansion

The amount by which unit length of a substance expands when its temperature is raised by one degree is called the ***coefficient of linear expansion*** of the substance. The temperature scale must be stated. Thus we can say that the coefficient of linear expansion of a solid is the fractional increase in length of unit length when its temperature is raised by one degree Celsius.

> **To calculate the increase in length of a body:**
>
> Increase in length = original length × coefficient of linear expansion × temperature rise
>
> This gives the *increase* in length. To find the *total* new length, the original length must be added.

For steel, the coefficient of linear expansion (denoted by the Greek letter α) is 0.000 012 per °C. Thus, a bar of steel 1 m long expands by 0.000 012 m for each °C rise in temperature; a 1 km bar expands by 0.000 012 km (12 mm) for each °C rise in temperature, and so on.

Some other typical values of the linear expansion coefficient, α, are:

Table 4.1 *Typical values of* α

Material	α (per °C)
Aluminium	0.000 023
Copper	0.000 017
Concrete	0.000 012
Steel	0.000 012
Invar steel	0.000 000 100
Common glass	0.000 009
Pyrex glass	0.000 003

4.1.2 Nickel-iron alloy (invar)

Invar is an alloy of iron and nickel (64 % iron, 36 % nickel) which has a coefficient of linear expansion of 0.000 000 1 per °C i.e., less than 1 per cent of that of steel. This is so small as to be negligible in most cases.

It is used for making measuring rods and tapes, watch and clock parts and other components which need must remain the same over a range of temperature.

Though there are many nickel-iron alloys, this very small coefficient of expansion applies only to this particular alloy which contains 36 % nickel.

Figure 4.1 Forth Road Bridge. (Photo: The Royal Commission on the Ancient Historical Monuments of Scotland)

4.1.3 Allowing for expansion in large metal structures

Large metal structures, such as bridges, often experience large variations in temperature, so allowance must be made for the linear expansion of the parts.

In large bridges, this expansion is quite large itself. For instance, the Forth Road Bridge is a steel structure with a total length (l) of about 1960 m. The maximum temperature range between winter and summer is -30°C to +30°C, a range of ΔT = 60°C. Using the formula above, the difference between the maximum and minimum lengths of the roadway would be:

$l \times \alpha \times \Delta T = 1960 \times 0\ 000\ 012 \times 60 = 1.41$ m

Figure 4.2 Main expansion joint at one of the main towers on the Forth Road Bridge.
(Photo: Forth Road Bridge Joint Board)

24 *Fire Service Manual*

Even a bridge with a span of 20 m could change by 14 mm between the hottest and coldest temperatures. Allowance for this expansion is often made by fixing one end of the bridge and resting the other on rollers, or on a sliding bearing, so that the bridge may expand and contract without exerting a side load on its piers. Railway lines used to be laid in 45 or 60 ft (13.7 or 18.3 m) lengths, with gaps to allow for expansion and contraction, but modern methods now make it possible for the expansion to be taken up as a tension or compression in the rail, with expansion joints at distances of about 800 m.

In buildings, the normal range of temperature is not usually so great, since internal heating maintains a reasonable temperature in winter and the building fabric protects steelwork from excessive external temperatures. Nevertheless, some allowance has to be made for expansion to prevent the steelwork from distorting the walls of the building, even if the allowance is only made by leaving a small clearance between the steel frame and the brickwork. In a fire situation, however, the increase in temperature may be very great and the situation could arise in which a long beam could exert sufficient side load on a wall and cause it to collapse.

Problems of expansion are also encountered with materials which are poor thermal conductors. In a fire situation, heating the inner face of a tall brick wall will cause that face to expand, while the outer face remains cool. This may cause the wall to lean outwards at the top and can result in collapse of the structure.

4.1.4 Thermostats

Imagine two strips of different metals with different coefficients of linear expansion, of the same length, laid side by side. Imagine then that the temperature of the surroundings increase. As the strips warm up, each will increase in length according to its own coefficient of linear expansion.

If the same two strips were fastened together throughout their length, an increase in temperature would cause them to distort into a curve, forming the arc of a circle. As the strip cooled, it would straighten out again. Such a strip is known as a bi-metallic strip. (Figure 4.3).

If one end of a bi-metallic strip is fixed, a change in temperature will cause the free end to move.

Figure 4.3 Bi-metallic thermostat

Left
Above a specified temperature, the bi-metal strip will bend. This causes the electrical contact to be broken which results in the current being switched off.

Right
A fire alarm using a bi-metallic strip.

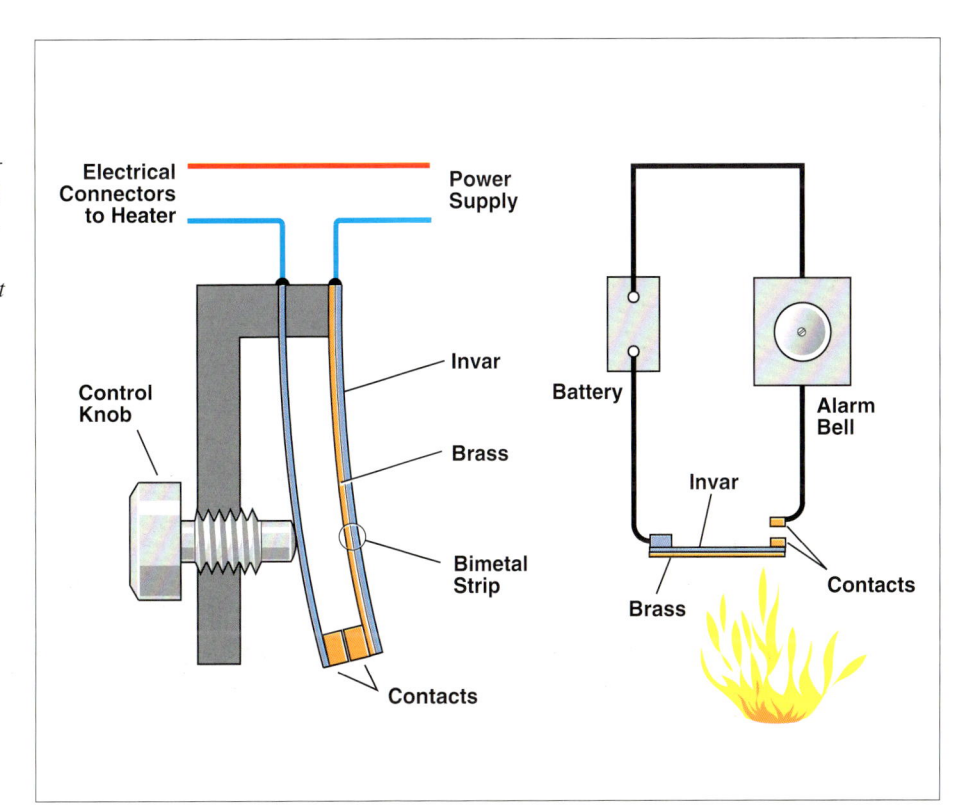

The movement of this free end can be made to open or close an electric circuit, to cause an alarm to be operated, or to switch off a heater. Such a device using a bi-metallic strip in this way is called a thermostat. (The principle is also used in rate-of-rise heat detectors.)

4.1.5 Coefficients of superficial and cubical expansion of solids

It can be shown mathematically that the coefficient of superficial (or area) expansion of a solid is twice the linear coefficient, and that of cubical expansion is three times the linear coefficient. Thus, the increase in volume ΔV of an object of volume V when the temperature is increased by ΔT is:

$$\Delta V = V \times 3\alpha \times \Delta T$$

so that the new volume will be $V + \Delta V$ (see Section 4.1.1 for the equivalent equation for linear expansion).

The expansion depends on the external dimensions of the solid and is not affected by any voids. The cubical expansion of a hollow metal box is the same as that of a solid block of the same metal of the same (external) volume as the box.

4.2 Thermal expansion of liquids

4.2.1 Cubical expansion

Since liquids have no definite shape and, therefore, no fixed dimensions other than volume, the only expansion which can be measured is that of cubical expansion.

Since a liquid has to be contained in a vessel, the apparent expansion of the liquid is affected by the expansion of the vessel, and the apparent expansion is, therefore, always less than the real expansion. However, the coefficient of cubical expansion of liquids is considerably greater than that of solids so (with the exception of water, which is dealt with below) the expansion of a liquid is always greater than that of its container.

Consider this comparison of the cubical expansion coefficients of glass, mercury and alcohol:

Material	Cubical expansion/°C)
Glass	0.000 024
Mercury	0.000 190
Alcohol	0.001 100

Thus, the thermal expansion of mercury is about 8 times that of glass, while that of alcohol is nearly 50 times that of glass. This is important in the design of thermometers.

The coefficient of cubical expansion of steel is 0.000 036/°C, and that of many liquids is of the order of 0 001/°C, i.e., about 30 times as much. Because of this, a sealed container (such as a storage tank) which is *completely* full of liquid may be a hazard in a fire situation (or even when exposed to strong sunlight) because of the internal pressures generated by expansion. If a pressure relief valve is fitted, this will allow the escape of liquid. The problem will be greatly reduced if the tank is not completely full, and there is an air space.

The so-called "frangible bulbs" used in many conventional sprinkler heads are sealed glass bulbs full of liquid. These break to operate and release water from the head when they are heated to a selected temperature, for example, when exposed to hot fire gases accumulating under the ceiling.

4.2.2 The effect of expansion on density

Since the density of a substance is the ratio of its mass to its volume, an increase of temperature results in a decrease of density; or conversely, the volume of a given mass of the substance increases as its temperature rises.

Water behaves in a peculiar way. Its expansion is not uniform: the expansion between 30°C and 50°C is double that between 10°C and 30°C. On cooling below 10°C, water contracts until its temperature reaches 4°C. On further cooling it expands until its volume at 0°C is 1.000 120 times greater than its volume at 4°C. It also expands further when it freezes. This means that water in ponds and lakes freezes from the top downward and, once the temperature on the surface has fallen to 4°C, further cooling of the lower level can only occur by conduction. This conduction is slow because water is a poor conductor.

4.3 The expansion of gases

4.3.1 Temperature, pressure, volume

Since a gas expands to fill all the available space, the volume of a gas may be changed by altering the volume of its container. If the volume is decreased, the pressure is increased.

This can be explained by saying that the same number of molecules of the gas occupy a smaller space and, therefore, collide with each other and with the container walls more frequently. The pressure is due to these collisions: more collisions, more pressure.

In a liquid, the molecules are much closer together to start with in than in a gas: the spacing of molecules in a liquid is comparable to that in a solid; though, because they are moving so quickly they do not remain in the regular structure of a solid. Because molecules in a liquid are so close together to start with, they cannot be compressed further. This is why gases can be compressed but liquids, generally, cannot.

Heating a gas increases the kinetic energy of the molecules which, therefore, move faster, and collide more frequently. So, heating a gas increases its pressure – provided its volume is unchanged. By increasing its volume as it is heated, the pressure can be kept constant.

We can see, then, that there are three variables which change with each other when dealing with a gas, namely **temperature**, **pressure** and **volume**. When dealing with a solid or a liquid, temperature and volume are important, but pressure is not so important.

4.3.2 The gas laws

There are three gas laws:

- Boyle's Law;
- Charles' Law; and
- The Law of Pressures.

These combine into the **General Gas Law**.

As we have seen, each solid or liquid will expand with rise in temperature by an extent determined by the coefficient of cubical expansion. However, all gases expand by the same amount for the same temperature rise.

Changes in volume of a gas depend on changes in temperature and pressure. To study the interaction between temperature, pressure and volume, one of these quantities is kept constant and the dependence of the other two on each other can then be studied. (Note: the mass of gas remains constant.)

This method of study provides the basis of the gas laws, the rules by which the behaviour of gases can be determined.

4.3.2.1 Boyle's Law

The change in volume of a gas caused by changes in pressure alone is the subject of the first of the gas laws, known as Boyle's Law. This states that:

> **Boyle's Law**
> **For a gas at constant temperature, the volume of a gas is inversely proportional to the pressure upon it.**

Experiments show that if the pressure applied to a given volume of gas is doubled, the volume is halved. If the pressure is trebled, the volume is reduced to one-third, provided the temperature is constant.

If we have a cylinder whose capacity is 1m^3 it can contain 1 m^3 of a gas at 1 atmosphere, but if 120 m^3 of gas at atmospheric pressure are compressed and pumped into the same cylinder, the pressure will be 120 atmospheres (atm). If half of the gas is allowed to escape, then the pressure will fall to 60 atm.

This is why the pressure gauge on a breathing apparatus set is a measure of how much air there is in the cylinder.

Figure 4.4 Breathing Apparatus Compressor.
(Photo: Hamworthy Compressor Systems Limited)

In practice, when a gas is compressed – for example, when a breathing apparatus cylinder is charged, or a tyre in pumped up – heat is generated and the temperature increases: the valve gets warm. This heat is only generated by the pumping operation. If the pressure of the gas is measured before the temperature has returned to its original level, Boyle's Law does not hold, since the temperature of the gas is not the same as it was before it was pumped in: this is why the **'constant temperature'** part of the law is important – the calculations do not work if it is not fulfilled.

Mathematically, if V_1 and P_1 are the initial volume and pressure, and V_2 and P_2 are the final volumes and pressure, then

$$\frac{V_2}{V_1} = \frac{P_1}{P_2}$$

$$P_1 V_1 = P_2 V_2$$

So,

| Initial pressure | × | Initial volume | = | Final pressure | × | Final volume |

4.3.2.2 Charles' Law

Experiments show that all gases expand by $1/273$ of their volume at 0°C for each 1°C rise in temperature, provided that they are maintained at constant pressure. Since the expansion for each 1°C rise in temperature is quite large, it is essential to take the initial volume at 0°C. These experiments were carried out by a French scientist named Charles at the beginning of the 19th century, and the law named after him states:

> **Charles' Law**
> **The volume of a given mass of gas at constant pressure increases by $1/273$ of its volume at 0°C for every 1°C rise in temperature.**

As gases expand, their density decreases and they become buoyant. This is why hot air rises, how hot air balloons work and why hot smoke and fire gases collect at the top of rooms.

It will be seen from Figure 3.3 (page 17), which shows the change of volume of a gas with temperature, that the Kelvin (or Absolute) scale of temperature must be used, and that the relationship between volume and temperature is:

$$\frac{V_1}{T_1} = \frac{V_2}{T_2}$$

where V_1 and T_1 are the initial volume and absolute temperature and V_2 and T_2 are the final volume and absolute temperature (the Kelvin temperature, **NOT** the Celsius temperature). In other words, the volume of a given mass of gas is directly proportional to its absolute temperature, provided that its pressure is kept constant.

4.3.2.3 The Law of Pressures

The previous two laws lead to a third law concerned with the relationship between the pressure

28 *Fire Service Manual*

> **The Law of Pressures**
> The pressure of a given mass of gas is directly proportional to its absolute temperature, provided that its volume is kept constant.

This is expressed mathematically as:

$$\frac{P_1}{T_1} = \frac{P_2}{T_2}$$

4.3.2.4 The General Gas Law

The three gas laws can be combined into a single mathematical expression:

$$\frac{P_1 V_1}{T_1} = \frac{P_2 V_2}{T_2}$$

This general expression may be used for a given mass of gas when pressure, temperature and volume all change.

Remembering this law will allow you to remember all three – you can just remove the quantity which is being kept constant from both sides of the equation. For example, if volume is being kept constant, remove V_1 from the left-hand side and V_2 from the right hand side (because they are equal, they cancel), and insert the pressure and temperature values that you know.

It is important to remember that these gas laws are applicable to all gases provided that they *remain* as gases over the temperature and pressure range involved. When the temperature and pressure reach levels at which the gas liquefies, the gas laws no longer apply.

Figure 4.5 Smoke layer at top of room.
(Photo: The Fire Experimental Unit)

and temperature of a gas when the volume is kept constant. This is the case when a cylinder of gas, whose valve is closed, is heated, as could happen if it were in a fire.

Figure 4.6 Fire involving liquefied petroleum gas tanks.
(Photo: HM Fire Service Inspectorate)

Physics and Chemistry for Firefighters

4.4 The liquefaction of gases

As has been noted previously, an increase in pressure raises the boiling point of a liquid.

Many substances which are gases at normal temperatures and atmospheric pressure can be compressed to such an extent that their boiling point is raised above atmospheric temperature and the gas liquefies (e.g., propane, ammonia).

Other gases cannot be liquefied at atmospheric temperature no matter how great a pressure is applied. These are the so-called 'permanent gases'. However, if the temperature is lowered sufficiently, it becomes possible to liquefy them by compression (e.g., methane, oxygen).

4.4.1 Critical temperature and pressure

For each gas, there is a **critical temperature** above which it cannot be liquefied by increasing the pressure alone. For example, carbon dioxide can be compressed to a liquid at 20°C, but at 40°C it will remain a gas.

Its critical temperature is in fact 31.1°C. Below this temperature it can be liquefied by increased pressure and it should properly be described as a vapour. Above this temperature it cannot be liquefied and is properly described as a gas, or, to emphasise the fact that it is above its critical temperature, **a 'true gas'**.

The pressure required to liquefy a vapour at its critical temperature is called the **critical pressure**.

Some typical values of critical temperatures and critical pressures are shown in Table 4.2.

4.4.2 Liquefied gases in cylinders

Many materials such as fuel gases are liquified under pressure and stored and transported in cylinders.

Liquefied gases in cylinders do not obey the gas laws, since, below the critical temperature, any change in temperature, pressure or volume will result in either the liquefaction of gas or the evaporation of liquid. Thus the pressure in a cylinder of liquefied gas will remain constant as gas is drawn off (provided the temperature remains constant) since more liquid will evaporate to make up for the gas drawn off until all the liquid is evaporated.

> **For cylinders of liquified gas, the cylinder pressure is not any indication of the amount of gas in the cylinder.**
>
> **A 'true gas' will obey the gas laws and the pressure will fall as gas is drawn off. Thus the pressure in the cylinder is an indication of the quantity of gas it contains.**

When liquefied gases are stored in cylinders, allowance must be made for expansion of the liquid in case the cylinder is heated beyond the critical temperature and the liquid turns into a vapour.

Table 4.2 *Critical temperatures, T_c and critical pressures, p_c for various substances*

	T_c (°C)	p_c (atm)
Water (steam)	374.0	-
Sulphur dioxide	157.0	219.0
Chlorine	144.0	78.0
Ammonia	132.0	77.7
Nitrous oxide	39.0	-
Carbon dioxide	31.1	73.1
Methane	-82.1	45.8
Oxygen	-119.0	50.0
Nitrogen	-147.0	33.7
Hydrogen	-240.0	12.9

This could lead to a substantial increase in pressure, with a risk of explosion. To minimise this danger, cylinders are never completely filled with liquid.

The amount of liquefied gas which may be charged into a cylinder is determined by its filling ratio, which varies from gas to gas and depends, among other things, on the density of the liquid.

$$\text{Filling ratio} = \frac{\text{Weight of liquefied gas which may be charged}}{\text{Weight of cylinder completely full of water}}$$

The filling ratio for ammonia is 0.5, so that a cylinder capable of holding 10 kg of water may only be charged with 5 kg of ammonia. A cylinder of the same size could be charged with 12.5 kg of sulphur dioxide, for which the filling ratio is 1.25.

4.5 Sublimation

In the laboratory it is possible to produce such low pressures that the boiling point of water can be reduced to 0°C and lower. When this happens, ice does not melt to form water, but will vaporise completely as the temperature rises.

This direct change from solid to vapour without forming an intermediate liquid is given the special name of **sublimation**.

In order to achieve sublimation with water, extremely low pressures are required, but solid carbon dioxide sublimes at atmospheric pressure.

At higher pressures, carbon dioxide shows the normal sequence of melting followed, at a higher temperature, by boiling, so that under pressure – and only under pressure – it is possible to have liquid carbon dioxide.

Physics and Chemistry for Firefighters

Chapter 5

Chapter 5 – Heat transmission

> As has been discussed earlier, heat always travels from high-temperature regions to low-temperature regions. In this Chapter, the ways in which heat energy can flow from hot bodies to cooler bodies will be discussed.

Heat energy always flows from regions of high temperature to regions of lower temperature. Heat will always flow when there is a temperature difference, no matter how small that temperature difference is.

There are three methods by which heat may be transmitted (Figure 5.1):

- conduction;
- convection; and
- radiation.

5.1 Conduction

Conduction may occur in solids, liquids or gases, although it is most clearly present in solids. In conduction, heat energy is passed on from each molecule to its nearest neighbour, with heat flowing away from the source of heat towards low temperature regions. The transfer of heat can be imagined to take place in much the same way as water in buckets being passed down a line of people in a 'bucket chain'. In the bucket chain each individual will only move a very small distance to either side of their mean position; it is only the water which passes on. In conduction of heat, the molecules vibrate about a mean position and pass on heat energy by colliding with their neighbours.

Thermal conductivity, the ability to conduct heat varies between materials. Most metals conduct heat relatively easily and are, therefore, classed as good conductors though the ability to conduct heat varies between metals.

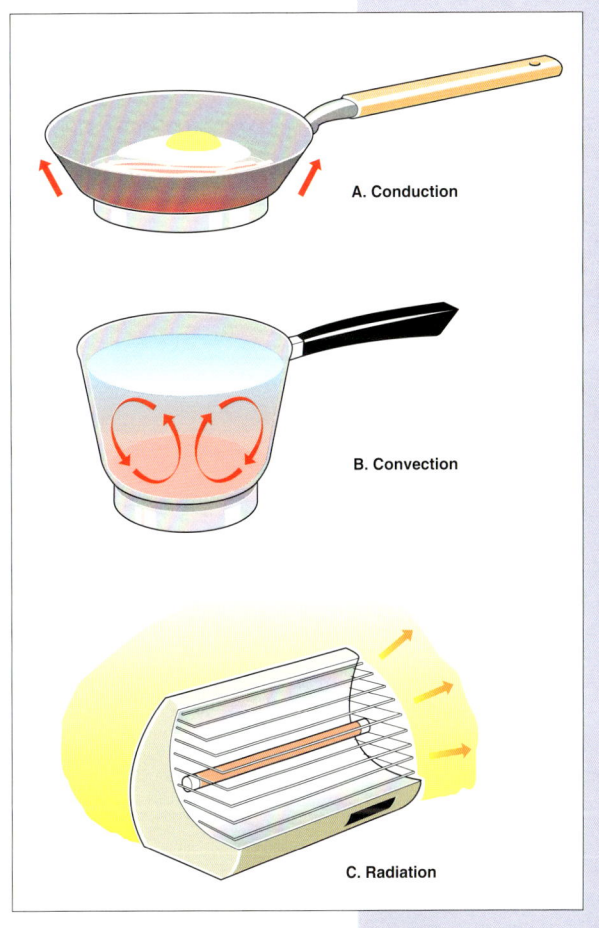

Figure 5.1 Diagram illustrating conduction, convection and radiation.

The best conductors of heat are silver and copper. Aluminium has about half the thermal conductivity of silver and iron about one-eighth. Non-metallic solids are poor conductors and, besides mercury, which is a metal, liquids and gases are very poor conductors of heat.

In fact, some solids and also liquids and gases are sometimes referred to as heat insulators because they are such poor conductors. In general good conductors of electricity (e.g., metals) are good conductors of heat, while poor conductors of electricity are good thermal insulators (e.g., most plastics).

Thermal conductivity can be measured experimentally and is usually denoted by the symbol K.

The flow of heat is measured in Joules per second (J/s) and this unit is the Watt (W). So:
$$1 \text{ J/s} = 1 \text{ W}$$

Thermal conductivity in the SI system of units is measured in Watts per metre per degree Kelvin (W/m K).

Thermal conductivity is important at most stages of a fire, but during the fully developed fire there is the danger of fire spread. As steel conducts heat very well, a steel girder passing through a fire wall may conduct sufficient heat through to the neighbouring compartment (room) to start a fire there. It is not necessary for flames to spread through the fire wall itself. (Figure 5.2).

Imagine a door built to separate rooms in case of a fire. If a fire occurs, a plain steel door will conduct heat rapidly to the other side, which could potentially cause the fire to spread outside the room.

Figure 5.2 Sketch showing how fire may be spread in a building due to the conduction of heat along an unprotected steel girder.

On the other hand, a wooden door, though it may burn, is initially a better barrier to heat as it is such a poor conductor. We can see then that the relative conductivity of building materials may be an important factor in the fire-resisting ability of a structure.

5.2 Convection

This occurs only in liquids and gases. It takes place for example, when a pan of water is heated (see Figure 5.3).

A pan full of water is heated from the bottom on a gas ring. As the water warms up, it expands and, therefore, becomes less dense, and so a given volume is lighter.

As the heated liquid is buoyant, it rises and colder, denser fluid takes its place at the bottom. This then becomes heated and so a circulation is set up. Heat energy is carried throughout the fluid by the molecules as they move until the water is the same temperature throughout. Compare this to conduction, in which the molecules do not move from their position: in convection it is the movement of the liquid or gas molecules through the mass of fluid which spread the heat energy around.

Convection is used in domestic hot water systems (Figure 5.4) and in many heating systems using so-called 'radiators'. Most of the heat from these radiators is in fact carried away by convection. It was also used in the 'thermo-syphon' system (now largely replaced by the pump-assisted system) of cooling motor engines.

Convection also causes the updraft in chimneys (see Figure 1.2). When a fire occurs in a building, convection currents can convey hot gases produced upwards through stairwells (Figure 5.5) and open lift and service shafts, thereby spreading the fire to the upper parts of buildings.

If the hot gas products escape from the upper levels, cool air must enter at low level to replaces them. This will, in addition, help to maintain the burning.

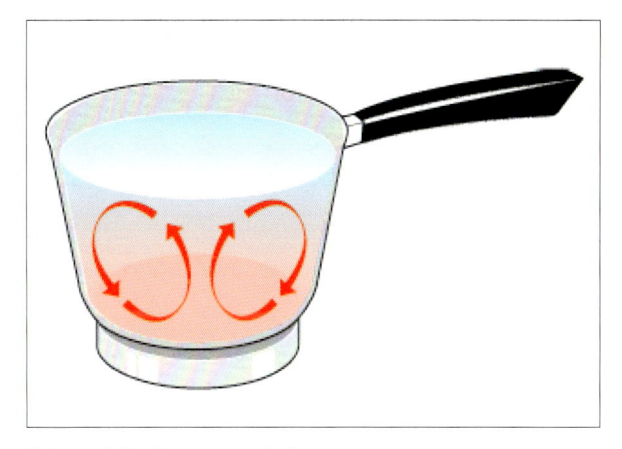

Figure 5.3 Convection in heated water.

Figure 5.4 Small bore heating and hot water system.

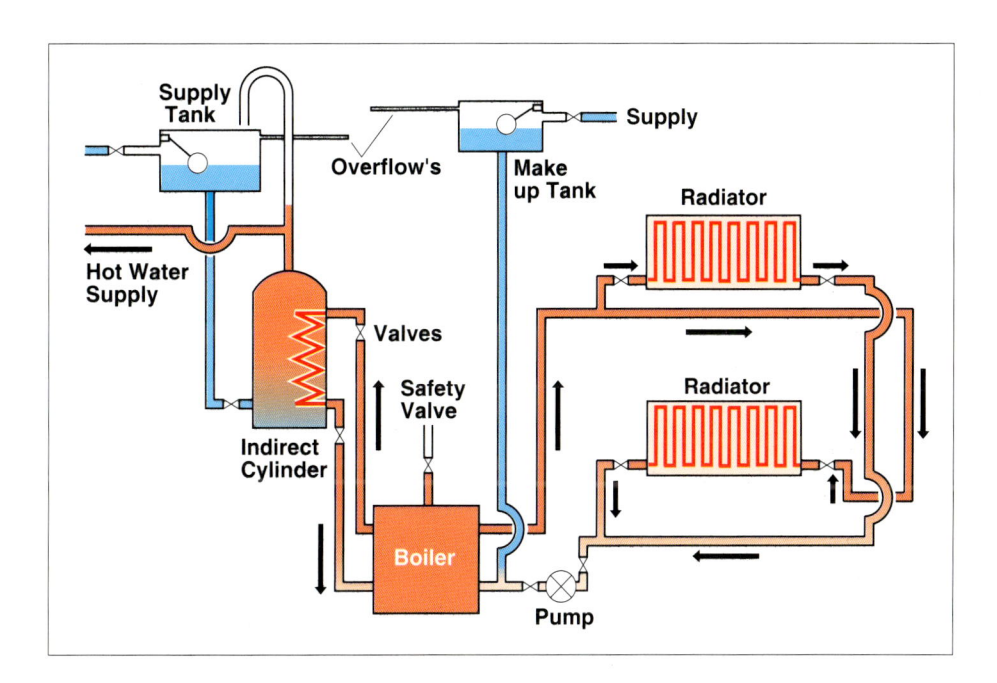

Physics and Chemistry for Firefighters

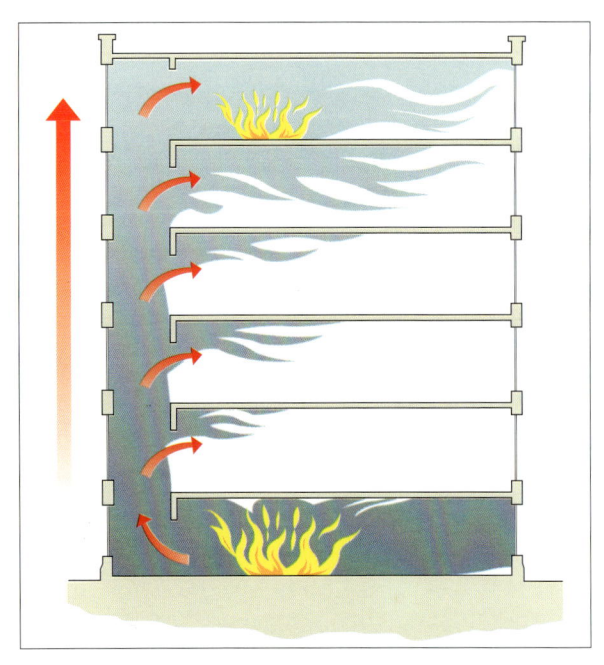

Figure 5.5 Sketch showing how fire on a lower floor can spread to upper floors by convection.

[The term "convective heat transfer" is used to describe the transfer of heat between a fluid (gas or liquid) and a solid. For example, a hot object in air loses heat partly by convection. The layer of air next to the hot surface becomes heated and, therefore, buoyant with respect to the surrounding cold air: it rises, carrying away the heat and is replaced by cold air. This in turn becomes heated, and a convection current is set up which cools the solid.]

5.3 Radiation

Heat may also be transmitted by a means which is neither conduction nor convection, nor requires an intervening medium. Energy from the sun passes through empty space to warm the earth. A radiant heater placed at high level in a room can be felt at lower levels, where neither conduction nor convection can carry it. This method of heat transmission is called *radiation* and does not involve any contact between the bodies which are providing and accepting the heat. To all intents and purposes, it behaves in the same way as light ("visible radiation") in that it travels in straight lines, will cast shadows, and will be transmitted through some materials and not others.

Heat is radiated as *infra-red* radiation, which is part of the spectrum of electromagnetic radiation. Radio waves, microwaves, visible light and X-rays are all part of this spectrum; the only thing which makes one form of radiation different from another is the wavelength of the radiation. The following table shows how the different forms of radiation occupy the electromagnetic spectrum.

Radiation	Wavelength (m)
Radio waves (UHF to long wave)	10^{-1} to 10^{4}
Microwave	10^{-3} to 10^{-1}
Infrared	8×10^{-7} to 10^{-3}
Visible light: from red	8×10^{-7} red
to violet	4×10^{-7} violet
Ultraviolet	4×10^{-7} to 10^{-8}
X-rays	10^{-8} to 10^{-13}
γ (gamma)-rays	less than 10^{-13}

Different parts of the electromagnetic spectrum have been given different names simply for convenience. What we call "visible light" is so-called because energy in the interval 8×10^{-7} to 4×10^{-7} m can be detected by the eye. "Infra-red" radiation cannot be detected by the eye as it is beyond the "red" end of the visible spectrum. It also contains less energy than visible radiation, which in turn is less energetic that ultra-violet radiation. UV radiation causes damage to biological systems, but only a small amount reaches the surface of the earth from the sun as it is absorbed by the ozone layer in the upper atmosphere.

All types of electromagnetic radiation produce a heating effect when they are absorbed by a body. This will depend on the amount of energy absorbed. A proportion of the energy radiated from the sun is radiated as visible light. If a body is heated above ambient temperature, it will radiate heat in the infra-red region of the spectrum. These "energy waves" have wavelengths longer than those of visible light.

All forms of electromagnetic radiation travel in straight lines at *the speed of light*, 3×10^{8} m/s.

Figure 5.6 Diagram showing the inverse square law as it applies to radiation.

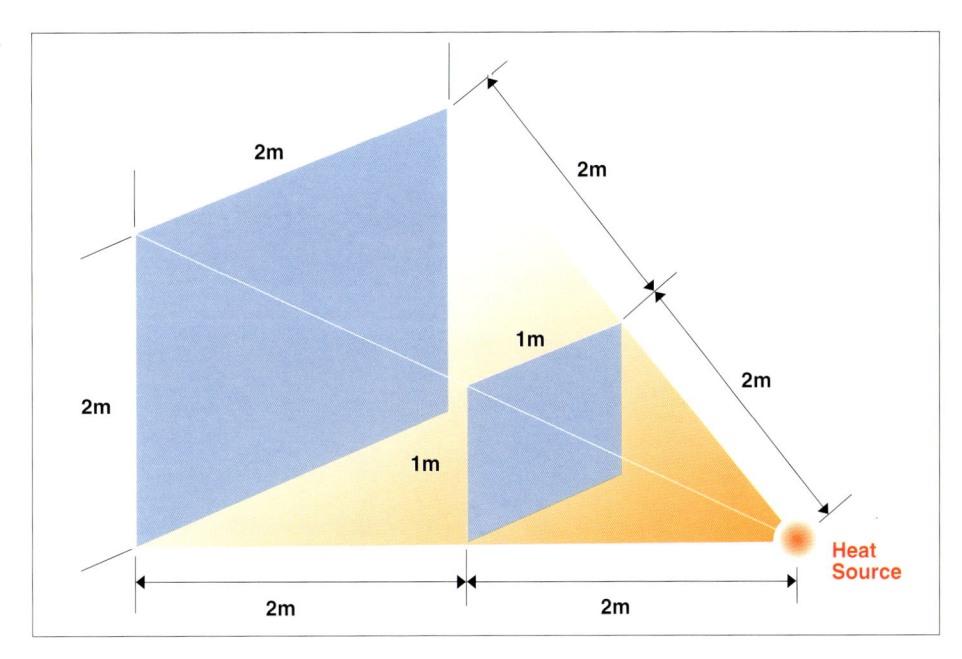

Radiation travels at this speed in a vacuum, but more slowly through matter such as air, water and glass.

The intensity of radiation – that is, how much energy reaches a surface of a given size – falls off inversely as the square of the distance from the source of radiation. This means that at twice the distance the intensity is one quarter; at three times the distance, the intensity is one-ninth, and so on. This inverse-square law is demonstrated in Figure 5.6.

The square with 1 metre sides placed at, say, 2 metres from the source will throw a shadow with 2 metre sides on a second sheet placed 4 metres from the source. Thus the energy falling on 1 m^2 is the same as that which would have fallen on an area of 2 m × 2 m = 4 m^2 at a distance of 4 m. So the energy per square metre at 4 m is one quarter that at 2 m, i.e. one quarter at twice the distance. This is important when considering the effect of radiation from a heat source such as a fire: a body of a given size and composition will heat up more slowly the further it is away from the radiation source.

When radiant energy (which, of course, includes infra-red radiation) falls on a body, there are three possibilities:

- **Transmission.** If energy passes through the body without warming it, it has been transmitted through the body. For example, 'transparent' materials transmit light.

- **Absorption.** The energy is absorbed by the body, whose temperature is raised; and

- **Reflection.** The energy may be reflected back from the surface in the way that light is from a shiny surface. Reflected energy does not enter the body, it just "bounces off" the surface.

Some substances absorb selectively: they allow some forms of radiation to pass, but not others. Glass, for example, allows light to pass but absorbs infra-red radiation – so glass may be used as a fire screen: the heat is stopped but the fire may be seen through it. However, for other reasons, such as its tendency to break under relatively low pressures and its behaviour at high temperatures, ordinary glass is not a good fire barrier.

Carbon dioxide and water vapour also exhibit this property. The sun's radiant energy, falling on the earth passes through the atmosphere and warms the ground, while the resulting infra-red radiation from the ground is absorbed by the atmosphere and so does not readily escape back into space. This is the cause of the 'Greenhouse Effect' – as industry, homes and transport have released more carbon dioxide into the atmosphere, the tendency to retain heat at the earth's surface has increased. It is believed that this may be producing a noticeable effect on the climate.

Physics and Chemistry for Firefighters

The ability of carbon dioxide and water vapour to absorb infra-red radiation in very narrow regions of the electromagnetic spectrum means that they also emit radiation in the same regions when they are hot. Radiation from small flames (e.g., at the early stage of a fire) is dominated by radiation from these gases, specifically at wavelengths of 2.8×10^{-7} and 4.4×10^{-7} m. Many infra-red detectors make use of this fact to distinguish between radiation from a flame and a hot solid object (such as a heating element) which emits over a wide range of wavelengths.

Some substances, such as pitch, transmit infra-red radiation, but absorb light.

The condition of the surface of a body affects its ability to absorb or reflect radiation. White or polished metal surfaces are the best reflectors and poor absorbers, while matt black surfaces are bad reflectors and good absorbers.

This is why white clothes, white-painted houses and cars, etc., are often used in hot climates. Snow and ice are poor absorbers of heat and reflect radiation very well. Because of this they melt slowly in direct sunlight. Melting will occur if the air temperature is raised, and heat is transferred by convection (Section 5.2).

Experiments have been carried out in which coal dust or other black powders have been spread on snow in order to accelerate melting. The sun's heat is more readily absorbed by the black powder, so its temperature rises and heat is transferred by conduction from the powder to the snow beneath.

Good reflectors of heat are also poor radiators. A polished silver teapot retains its heat better than a blackened teapot, in spite of silver being a good conductor. For the so-called radiators of a hot water system to radiate effectively they should be painted black, not a light colour as is usually the case. However, the principal way by which they warm a room is by heating the air immediately around them, which then rises, producing convection currents which spread heat throughout the room. The other side of the coin is that the heater is being cooled by convection.

Many fires have been caused by radiation – one of the most common is clothing being ignited by being placed too close to a source of radiation (Figure 5.7), as sometimes happens when people air clothes on a clothes horse placed too near a fire. Radiant heat from the sun passing through a glass window has sometimes been concentrated by an object inside the house which acts as a lens, such as a magnifying glass or a shaving mirror. The old-fashioned "bottle glass" used in windows will also concentrate the sun's rays and could, in principle, cause a fire, but a bottle, or pieces of broken bottle cannot.

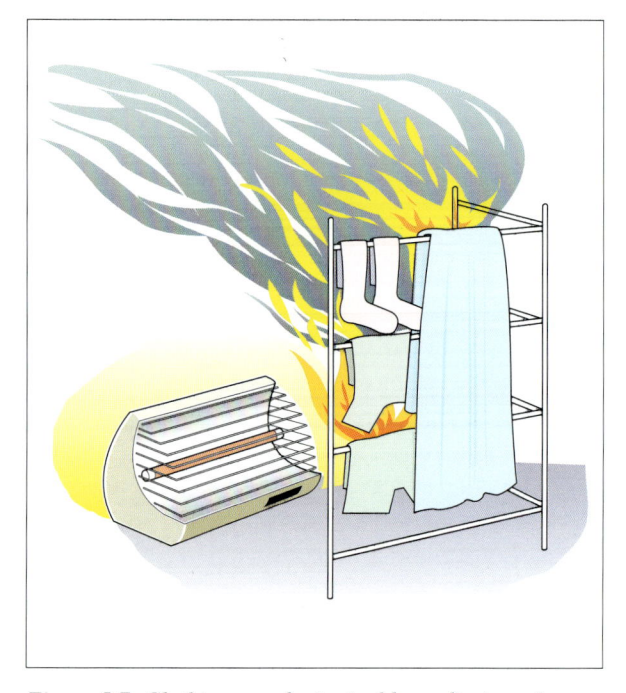

Figure 5.7 Clothing may be ignited by radiation when placed too close to a source of radiated heat.

Physics and Chemistry for Firefighters

Chapter 6

Chapter 6 – The Basis of Chemistry

The Chemistry of Combustion

Up to now, we have considered the physical properties of matter and heat, the properties that decide how bodies will behave when energy is supplied to them.

In the rest of this volume we will deal with the chemistry of combustion – the reactions by which energy is released in fires. Before discussing combustion in detail, it is necessary to talk about understand some of the basic concepts of chemistry.

Chemistry is a complicated subject bristling with long and difficult names to pronounce, and with intricate formulae used by the chemist. There are, of course, many text books available to the student on chemistry and, in presenting an opening to the study of fire-fighting techniques, it is difficult to decide exactly how much should be included. Many new processes and materials have become available in recent years. Firefighters are faced with so many new substances, particularly new building materials, during the course of their work, that they *must* have some idea how they will react when involved in fire. The particular hazards of many flammable materials and chemicals are dealt with separately in Parts 6b and 6c of the Manual of Firemanship. However, in this Chapter it is proposed to deal with those aspects of chemistry which apply to the study of fire techniques, and to lead on to discuss some of the more hazardous chemical substances from a purely chemical point of view.

6.1 The basis of chemistry

Chemistry is the science of the composition of substances, their properties and reactions with each other. Substances may be solids, liquids or gases, in living or non-living systems, but all have one common factor – they consist of chemicals.

6.2 Atoms and molecules

Chemists recognise two distinct classes of substances; those which consist of a single chemical (elements and compounds) and those which are mixtures. A mixture may be separated into its constituents by some physical or mechanical means; for example, a mixture of salt and sand can be separated by dissolving the salt in water, leaving the sand behind. But to separate or change a single chemical substance, a chemical reaction is required.

Whether the substance is single or a mixture, it is made up from many millions of very tiny particles which the chemist calls molecules. (See Section 1.5) A mixture will contain more than one type of molecule, whereas a chemical compound contains only one type of molecule. Molecules of the same substance are all exactly alike in their properties and behaviour.

A molecule can be said to be the smallest particle of a compound capable of existing independently. The common substance chalk occurs in large quantities and in many different forms. For example, it is found in cliffs as lumps, or as a powder; it is, nevertheless, always recognisable as the same material, known chemically as calcium carbonate. This material is formed from innumerable calcium carbonate molecules. Each molecule is composed of even smaller particles called atoms. Every calcium carbonate molecule is exactly the same; each contains five atoms.

Molecules are formed from atoms. The number of different atoms comprising their molecules is relatively small. The molecules of all substances comprise various combinations of atoms, from approximately 90 different types of atom.

Atoms are the 'building blocks' of all substances. Unlike molecules, which can be broken down or changed during chemical reactions, atoms cannot be split *chemically* into anything smaller[1]. During chemical reactions the atoms rearrange to form different molecules, but the atoms themselves remain the same. They are the smallest particles to take part in chemical changes. Atoms are extremely small, their diameters being

between $\dfrac{1}{10000000}$ mm and $\dfrac{4}{10000000}$ mm

Substances formed entirely from one type of atom are called elements. There is an element corresponding to each different type of atom. Thus carbon, being formed entirely from carbon atoms, is an element. Similarly iron, containing only iron atoms, is another element. Elements may be composed of molecules made up from identical atoms joined together, or they may be composed of single atoms. The element oxygen consists of oxygen molecules, each molecule being two oxygen atoms joined together, whereas the element magnesium consists of single magnesium atoms. When we again consider the molecule of calcium carbonate (chalk) we find that it is composed of one atom of the element calcium, one atom of the element carbon, and three atoms of the element oxygen, all chemically bound together. A list of the names of some elements is given in Table 6.1; a full list is given in Appendix B.

6.2.1 Compounds and mixtures

When two or more atoms of different elements are chemically bound together to form molecules, all exactly the same, the substance formed is called a chemical compound. For example, each molecule of the compound calcium carbonate contains five atoms chemically bound together (one of calcium, one of carbon and three of oxygen). The compound formed from identical molecules can only be broken down or changed by a rearrangement of the atoms, known as a chemical reaction. A mixture (formed from two or more different sorts of molecules) can be separated by physical or mechanical means into the substances which make up the mixture.

Table 6.1 *List of some elements with their atomic number atomic mass and valency*

Name of element	Symbol	Atomic number	Atomic mass	Valency
Aluminium	Al	13	27.0	3
Bromine	Br	35	80.0	1
Calcium	Ca	20	40.0	2
Carbon	C	6	12.0	4
Chlorine	Cl	17	35.5	1, 3, 5, 7
Copper	Cu	29	63.5	1, 2
Fluorine	F	9	19.0	1
Gold	Au	79	197.0	1, 3
Helium	He	2	4.0	0
Hydrogen	H	1	1.0	1
Iodine	I	53	127.0	1, 3, 5, 7
Iron	Fe	26	56.0	2, 3
Magnesium	Mg	12	24.0	2
Mercury	Hg	80	201.0	1, 2
Neon	Ne	10	20.0	0
Nitrogen	N	7	14.0	3
Oxygen	O	8	16.0	2
Phosphorus	P	15	31.0	3
Potassium	K	19	39.0	1
Silicon	Si	14	28.0	4
Silver	Ag	47	108.0	1
Sodium	Na	11	23.0	1
Sulphur	S	16	32.0	2, 4, 6
Uranium	U	92	238.0	4, 6

---

1 They can be split by *nuclear* processes, which will be discussed elsewhere.

> **Atoms are the smallest particles.**
>
> **Molecules are atoms chemically bound together.**
>
> **Elements contain only one sort of atom, either alone or grouped into molecules containing only one type of atom.**
>
> **Compounds are formed from one type of molecule which contains more than one kind of atom.**
>
> **Mixtures contain more than one kind of molecule.**

6.3 Symbols

Chemical symbols are used as a way of describing chemicals in terms of formulae, which are complete descriptions of molecules in terms of the constituent atoms. Symbols give as much information as possible, whilst still being simple and quick to use. Formulae may also be used to describe the way that atoms in a molecule are grouped together. This information may give clues to how one chemical compound may react with another.

Every element is assigned a symbol (see Table 6.1), which is different from that of all the other elements. A symbol may be one letter or two; in the latter case the convention is to write the second letter as a small letter. Thus the symbol for nickel is written Ni and not NI. NI would be interpreted as a molecule containing one nitrogen atom (N) and one iodine atom (I) (such a compound does not exist). In many cases the symbols are the first letter of the name of the element, often followed by a second letter taken from that name. However, there are several common elements whose symbols bear no relationship to their modern names, since they are based on the old Latin or Greek names. For example, the element sodium has the symbol Na which is derived from the Latin natrium, and lead has the symbol Pb, derived from plumbum (hence plumber, and plumb line).

6.3.1 Using symbols to write formulae

When a symbol is written it represents one atom of the element. Thus: H represents one atom of hydrogen; O represents one atom of oxygen.

A formula always represents one molecule of the substance and shows which atoms are present in the molecule and how many of them there are. Thus H_2O represents one molecule of water, containing two atoms of hydrogen and one atom of oxygen, bound together chemically. Similarly carbon dioxide has the formula CO_2, representing the molecule which contains one atom of carbon and two atoms of oxygen. If a molecule contains more than one atom of the same type, the number of similar atoms is written at the bottom right of the appropriate symbol:

Calcium carbonate
$CaCO_3$ 1 atom of calcium, one atom of carbon and 3 atoms of oxygen

Phosphorous pentoxide
PO_5 1 atom of phosphorus and 5 atoms of oxygen.

To represent more than one molecule we write a number in front of the formula: thus three molecules of water are represented by $3H_2O$. This group of three water molecules contain six hydrogen atoms and three oxygen atoms.

$2MgO$ represents two molecules of magnesium oxide (and, therefore, a total of two magnesium atoms and two oxygen atoms).

6.3.2 Radicals

Certain groups of atoms, common to families of related compounds, are known as radicals. **A radical can be defined as: 'a group of atoms present in a series of compounds which maintains its identity regardless of chemical changes which affect the rest of the molecule'.**

Physics and Chemistry for Firefighters

To show these radicals, formulae are often written with brackets enclosing the radical and with a number beyond the bracket to indicate how many of these radicals are in the formula.

Radicals are not complete molecules and have no independent existence. For example, the formula of one molecule of calcium hydroxide (slaked lime) is Ca(OH)$_2$, indicating that it contains one calcium atom and two hydroxyl (OH) radicals. The molecule contains two oxygen atoms and two hydrogen atoms, but they are always paired, as OH. Another common radical is NO$_3$, the nitrate radical. The formula for aluminium nitrate is written Al(NO$_3$)$_3$, indicating that the trivalent aluminium atom is combined with three monovalent nitrate radicals. Schematically:

$$Al \begin{cases} NO_3 \\ NO_3 \\ NO_3 \end{cases}$$

The meaning of valency is discussed in Section 6.6. A list of common radicals is given in Table 6.2.

6.4 Atomic mass

The mass of one atom or one molecule is extremely small – of the order of 10^{-22} grams. It is of little practical value to quote the actual masses of atoms, but because atoms of different elements contain different numbers of protons and neutrons, knowing the mass is a big step towards identifying which element the atom belongs to.

It is, therefore, important to know how heavy one atom is in comparison with any other. The chemist, therefore, uses a relative atomic mass scale and not the actual masses of the atoms. Various scales have been proposed and, for technical reasons, the one most generally used is based on oxygen, which is given the atomic mass of 16.000. On this scale hydrogen has an atomic mass of 1.008. However, for normal purposes, the atomic masses can be rounded off, making hydrogen equal to 1.

We can then compare other atoms with hydrogen to see how many times heavier they are, so that we have the definition:

$$\text{Atomic mass} = \frac{\text{The mass of one atom of the element}}{\text{the mass of one atom of hydrogen}}$$

Table 6.2 *A list of common radicals*

	Name	Symbol
Valency 1	Ammonium	NH$_4$
	Bicarbonate	HCO$_3$
	Bromide	Br
	Chlorate	ClO$_3$
	Chloride	Cl
	Cyanide	CN
	Hydroxide	OH
	Iodine	I
	Nitrate	NO$_3$
	Nitrite	NO$_2$
	Perchlorate	ClO$_4$
Valency 2	Carbonate	CO$_3$
	Sulphate	SO$_4$
	Sulphide	S
	Sulphite	SO$_3$
Valency 3	Phosphate	PO$_4$

For example, the atomic mass of sodium is 23 (written Na = 23), meaning that an atom of sodium is 23 times heavier than an atom of hydrogen.

6.5 Molecular mass

In the same way, molecular mass is the mass of one molecule of the substance compared to the mass of one atom of hydrogen. For example, the molecular mass of water is 18 which means that one molecule of water is 18 times as heavy as one atom of hydrogen. Since a molecule consists of atoms joined together, the mass of the molecule is the sum of the masses of its component atoms. The molecular mass is found by adding together the atomic masses of those atoms present, thus: The molecular mass of sulphur dioxide (SO$_2$) is 64.

Molecular Mass = atomic mass of sulphur +
　　　　　　　　2 × atomic mass of oxygen
　　　　　　　= 32 + (2 × 16)
　　　　　　　= 64

Similarly: nitric acid HNO$_3$, molecular mass 63;

Mol mass	=	atomic mass H	+	atomic mass N	+	3 × atomic mass O
	=	1	+	14	+	(3 × 16)
	=	63				

6.6　Valency

When atoms combine to form molecules they do so in definite fixed ratios. For example, one sodium (Na) atom always combines with one chlorine (Cl) atom to give NaCl (common salt), but one magnesium (Mg) atom combines with two chlorine (Cl) atoms to give MgCl$_2$ (magnesium chloride). The 'combining power' of an atom depends on the arrangement and number of its electrons, but the mechanism of this is too complicated to discuss easily here.

The **valency** of an atom tells us how many chemical bonds the particular atom, or group of atoms (radicals) will form. Valencies are given in Table 3. When molecules are formed, the atoms or radicals generally combine in ratios in which the valencies are balanced. This property enables us to work out the correct formulae of many chemical compounds. For example, in magnesium oxide, Mg has a valancy of 2, and O has a valency of 2. To balance the valencies we need one Mg atom and one O atom; hence the formula MgO.

In potassium carbonate, potassium (K) has a valency of 1, while the carbonate radical has a valency of 2. Potassium carbonate requires two K atoms to combine with one carbonate radical, thus the formula is K$_2$CO$_3$.

In aluminium sulphate Al has a valency of 3, the sulphate radical has a valency of 2. To form the compound, two Al atoms are required to balance three SO$_4$ radicals (total number of "bonds" = 6), so that the formula for aluminium sulphate is Al$_2$(SO$_4$)$_3$.

6.6.1 Multiple valency

Several elements show more than one valency, including iron (Fe), copper (Cu) and nickel (Ni). The valency that the element shows depends on the particular circumstances – the other elements with which the element is combined, as well as on the conditions under which the reaction in which the compound is formed is carried out. More detailed knowledge of the chemistry of the element is required to predict which valency will be shown in any particular reaction. However, the names of the compounds formed are often adapted to help in deciding which valency state the element is in, in that particular compound, and hence to determine the correct formula.

6.6.2 Nomenclature

(1) -OUS and -IC
-OUS and -IC are used where an element shows two valencies.
-OUS always indicates the lower and -IC the higher valency. For example:
　Iron
　Ferrous: valency 2, e.g., FeCl$_2$ ferrous chloride.
　Ferric: valency 3, e.g., FeCl$_3$ ferric chloride.
　Tin
　Stannous: valency 2, e.g., SnBr$_2$ stannous bromide.
　Stannic: valency 4, e.g., SnBr$_4$ stannic bromide.

(2) Use of Roman numerals
A modern approach to the problem of multiple valency is to indicate which valency is being used by inserting the appropriate Roman numeral after the name or symbol of the element concerned, e.g.,
Iron (II) chloride Fe(II)Cl$_2$,
Iron (III) chloride Fe(III)Cl$_3$,
Tin (II) bromide Sn(II)Br$_2$,
Tin (IV) bromide Sn(IV)Br$_4$.

(3) -IDE
-IDE is used to indicate that a compound is made up of two elements only. By convention, metals are written before non-metals in names and formulae, thus:
　Magnesium oxide MgO
　(Mg valency 2; O valency 2).

　Potassium sulphide K$_2$S
　(K valency 1; S valency 2)

Physics and Chemistry for Firefighters

-IDE is also used exceptionally for a few radicals, e.g., -OH hydroxide. Thus:
Calcium hydroxide Ca(OH)$_2$
(Ca valency 2; OH valency 1).

(4) -ITE and -ATE
-ITE and -ATE are used where a compound contains more than two elements, one of which is oxygen. For two related compounds, that named -ITE always contains less oxygen than that named -ATE.

Sodium sulphite Na$_2$SO$_3$
(Na valency 1; SO$_3$ valency 2).

Sodium sulphate Na$_2$SO$_4$
(Na valency 1; SO$_4$ valency 2).

Potassium nitrite KNO$_2$
(K valency 1; NO$_2$ valency 1).

Potassium nitrate KNO$_3$
(K valency 1; NO$_3$ valency 1).

The ending -ITE and -ATE are related to the ending -OUS and -IC where the latter are used in the names of acids. -OUS leads to -ITE and -IC to -ATE. For example:

Sulphurous acid H$_2$SO$_3$ gives sulphites -SO$_3$
Sulphuric acid   H$_2$SO$_4$ gives sulphates -SO$_4$
Nitrous acid   HNO$_2$ gives nitrites   -NO$_2$
Nitric acid   HNO$_3$ gives nitrates   -NO$_3$

(5) Mono-, Di-, Tri-, Tetra-, Penta
Mono-, di-, tri-, tetra- and penta- are used in names to tell how many of a particular atom or radical are present.
Mono- 1 e.g., carbon monoxide CO,
Di-   2 e.g., carbon dioxide CO$_2$,
Tri- 3 e.g., sulphur trioxide SO$_3$,
Tetra- 4 e.g., carbon tetrachloride CCl$_4$,
Penta- 5 e.g., phosphorus pentachloride PCl$_5$.

(6) Per- always denotes that there is more oxygen present in the compound than would normally be the case:
Hydrogen oxide (water) H$_2$O,
Hydrogen peroxide H$_2$O$_2$,
Sodium chlorate NaClO$_3$,
Sodium perchlorate NaClO$_4$.

6.7 Simple equations

Consider a simple chemical reaction. When sulphur (a yellow solid element) burns in air, it combines with oxygen from the air, producing a colourless gas with a pungent choking smell. This gas is called sulphur dioxide (formula SO$_2$).

This can be stated simply as: "sulphur reacts with oxygen to form sulphur dioxide". A further simplification can be made by replacing 'reacts with' by "+" and 'to form' by an equals sign, "=". We then have:

Sulphur + oxygen = sulphur dioxide
     reacts with   to form

This statement can be simplified even further by replacing the names of the chemicals by symbols and formulae. (The molecules of oxygen, like those of most common elements that are gases, contain two atoms, but sulphur, like other solid elements, is assumed to consist of single atoms.) This gives:

$$S + O_2 = SO_2$$

This final statement represents the chemical equation for this reaction. It tells us that every sulphur atom involved reacts with one oxygen molecule to form one sulphur dioxide molecule.

> **It should be noticed that each side of the equation contains the same number of each type of atom present. This must always be the case since a chemical reaction involves only a rearrangement of atoms – atoms do not appear or disappear – the equation must 'balance'.**

Consider this other example. Magnesium (a metal) burns in oxygen to form magnesium oxide (a white powder). Magnesium reacts with oxygen to form magnesium oxide:

Magnesium + oxygen = magnesium oxide

$Mg + O_2 = MgO$

In this case, although there is the same number of magnesium atoms on each side of the equation, this is not true in the case of the oxygen atoms, where there are two on the left hand side and only one on the right hand side. This implies that oxygen atoms disappear during the reaction. The equation must be balanced before the equation is of any practical use – before it can tell us how the elements combine.

The equation above can be balanced by placing two molecules of magnesium oxide on the right hand side, thus:

$Mg + O_2 = 2MgO$

We now have equal numbers of oxygen atoms on each side of the equation, but the magnesium is 'out of balance'. By having two atoms of magnesium on the left hand side (instead of one):

$2Mg + O_2 = 2MgO$

the equation is now correctly balanced. Each side now contains equal numbers of each type of atom involved. It is of course possible to balance the equation by cheating. The equation:

$Mg + O_2 = MgO$

"balances" if we change the formula of magnesium oxide, thus:

$Mg + O_2 = MgO_2$

but MgO_2 does not exist. A chemical equation can only be balanced by changing the number of molecules present, not their formulae.

6.8 Use of chemical equations

The balanced equation: $2Mg + O_2 = 2MgO$ tells us that two atoms of magnesium react with one molecule of oxygen to produce two molecules of magnesium oxide.

These atoms and molecules have masses which are expressed in terms of their atomic and molecular masses (as described in Section 6.3). The atomic mass of magnesium is 24, that of oxygen is 16, and if we use this information together with the equation, we obtain:

2 Mg	+	O2	=	2MgO
2 x 24		2 x 16		2 (24 + 16)
48 units		32 units		80 units
Two magnesium atoms		Two oxygen atoms		Two MgO molecules each containing 1 magnesium and 1 oxygen atom

The 'units' are mass units, where one unit represents the mass of one hydrogen atom. Therefore, according to the equation, 48 units of magnesium will react with 32 units of oxygen to form 80 units of magnesium oxide. In other words the ratio:

$$\frac{\text{mass of magnesium}}{\text{mass of oxygen}} = \frac{48}{32}$$

An actual reaction between magnesium and oxygen will obviously involve millions of molecules of each substance. Suppose we 'scale up' this reaction until we have two million magnesium atoms instead of two. Then this number of magnesium atoms will react with one million oxygen atoms.

Two atoms of magnesium weigh 48 units, therefore, two million atoms of magnesium weigh 48 000 000 units.

Similarly, one molecule of oxygen weighs 32 units; therefore, one million molecules of oxygen weigh 32 000 000 units, so that the ratio:

$$\frac{\text{mass of magnesium}}{\text{mass of oxygen}} \text{ will again be } \frac{48}{32}$$

No matter to what extent the amounts of magnesium and oxygen are scaled up, the ratio will always be $^{48}/_{32}$. Therefore:

if 48 grams of magnesium are used, 32 grams of oxygen are needed. If 48 kg of magnesium are used, 32 kg of oxygen are needed, and 80g or 80 kg of magnesium oxide will be produced. For any other mass of magnesium, the masses of oxygen needed and magnesium oxide produced can be found by simple proportion.

6.9 Limitations of chemical equations

6.9.1 Reality

A chemical equation must be a summary of a known chemical reaction. For instance it is perfectly possible to write down the equation:

$$Cu + 2HNO_3 = Cu(NO_3)_2 + H_2$$
copper + nitric acid = copper nitrate + hydrogen

but such an equation is useless because it is found that when copper is placed in nitric acid, hydrogen is never produced. Therefore, the equation is not 'telling the truth', even though the sides balance.

6.9.2 Physical state

The equations we have been considering contain no information about the physical state of the chemicals, whether they are solids, liquids or gases, whether they are pure substances or are dissolved in water or some other solvent, or whether the solutions are dilute or concentrated. Sometimes it is important to specify the physical state. For example, the reaction of hydrogen and oxygen to form water vapour is associated with the release of a certain amount of heat. If the water vapour is allowed to condense, the latent heat of vaporisation is released. Thus, if we are quoting the amount of heat released by the reaction of hydrogen and oxygen, the physical form of the water must be stated. This can be done as follows:

$$2H_2 + O_2 = 2H_2O(g)$$
$$2H_2 + O_2 = 2H_2O(l)$$

where g and l refer to the gaseous (vapour) and liquid states respectively. As written, the second reaction releases more heat then the first, by an amount equal to the latent heat of evaporation of water.

6.9.3 Reaction conditions

Equations say nothing about the reaction conditions; whether heat must be used or pressure applied.

6.9.4 Heat

Equations do not tell us whether heat is given out or absorbed during a chemical reaction.

6.9.5 Rate of reaction

Equations say nothing about the rate of the reaction; whether it is slow, fast or inherently violent; or whether or not a catalyst[2] is necessary to make the reaction occur at a reasonable rate.

---

[2] A catalyst is a substance that alters the rate of a chemical reaction, but does not itself undergo a chemical change.

Physics and Chemistry for Firefighters

Chapter 7

Chapter 7 – Combustion

> Flames are so much part of our everyday experience that it seems strange that there is anything to write about regarding their nature. Certainly, every firefighter knows what dangers flames present, sees how they can spread and is interested in how they can be extinguished. However, many people, if asked what a flame actually is, would find it difficult to produce an answer. This section aims to shed light on the nature of flames, to give the reader an insight into their make-up and, especially, to highlight the different types of flame.

7.1 The Fire Triangle

For combustion or burning to occur, oxygen, usually from the air, must combine with a fuel. A fuel may be in any one of the three states (gas, liquid or solid) initially, but for flaming combustion to occur, a solid or liquid fuel must be converted into a vapour, which then mixes with air and reacts with oxygen. Smouldering combustion, on the other hand, involves a reaction between oxygen (from the air) and the surface of the fuel: this is a complex process and in general only occurs with solid fuels which char on heating.

A flame is a region in which a sustained, heat-releasing reaction between a fuel in the vapour state and oxygen takes place. This region also emits light, usually with a strong yellow colour, though there are substances such as methanol which burn with a weak blue flame which cannot be seen in strong light.

One way of discussing burning is in terms of the triangle of combustion (Figure 7.1). For combustion to occur three things are necessary: heat, oxygen and fuel. Combustion will continue as long as these three factors are present. Removing one of them leads to the collapse of the triangle and combustion stops.

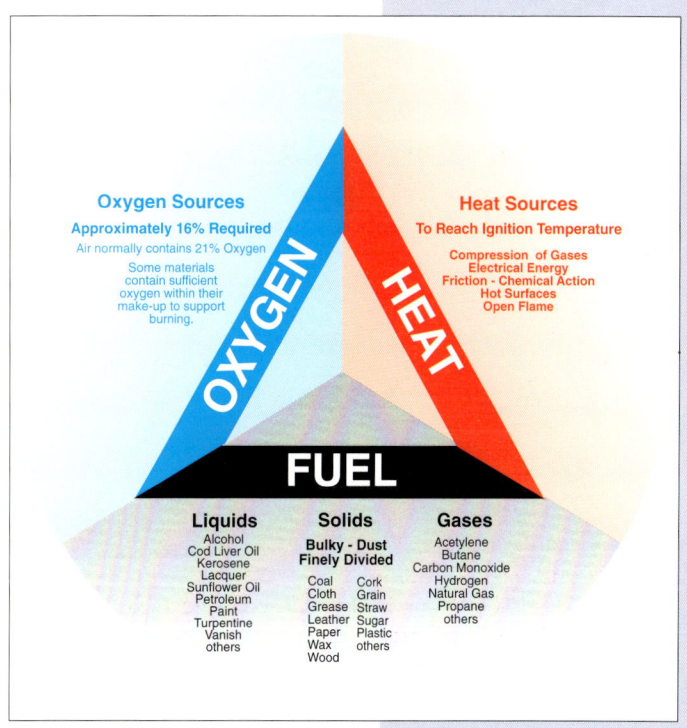

Figure 7.1 The triangle of combustion.

Physics and Chemistry for Firefighters   47

7.2 Heat of reaction and calorific value

All combustion reactions release heat energy and are, therefore, called **exothermic** reactions. The quantity of heat produced per unit weight of fuel can be calculated and is known as the **calorific value** of the fuel. For example, when 12 grams of carbon (the "gram atomic mass") are burned to carbon dioxide 392 920 Joules of heat are produced. This is the "heat of combustion", as tabulated in textbooks and handbooks: it refers to a standard amount of the fuel (the "mole") and has units kJ/mol. The calorific value for carbon is then:

$$\frac{392920}{12} = 32\,743 \text{ Joules per gram (J/g)}$$

as one "mole" of carbon contains 12 g.

Besides calorific value, the rate of heat release is also important. For example, burning magnesium produces less heat than the burning of carbon, but when rates of reaction are considered, we find magnesium has a much higher rate of combustion than carbon so that the heat is released much more rapidly. **Heat release rate is now considered to be a major factor in whether a fire will spread over materials**, and a device called a Cone Calorimeter has been developed to measure this quantity for wall linings and building materials in general as part of the assessment of their flammability and suitability for their intended use.

7.2.1 Oxidation

An oxidation reaction is a reaction which involves combination with oxygen or other oxidising agents. The following reactions are typical examples of combustion:

(i) The oxygen may be supplied by the air.

$$2C + O_2 \rightarrow 2CO \text{ (carbon monoxide)}$$
$$2CO + O_2 \rightarrow 2CO_2$$
$$2H_2 + O_2 \rightarrow 2H_2O$$

(note that the oxidation of C to CO is not a flaming reaction: the others are).

(ii) The combustion may take place using oxygen which is contained within the burning material, the combustible material and the supporter of combustion being together in the same compound:

$$4C_3H_5(NO_3)_3 \rightarrow 12CO_2 + 10H_2O + 6N_2 + O_2$$
nitroglycerine

(iii) Oxygen may be provided by one of the materials in a mixture of compounds. The 'thermite reaction' illustrates this principle:

$$Fe_2O_3 + 2Al \rightarrow Al_2O_3 + 2Fe + \text{heat}$$
thermite mixture

(iv) Elements other than oxygen may be considered as oxidising agents; examples of these are chlorine and fluorine. "Combustion" may occur with these substances; for example, hydrogen will burn explosively with chlorine:

$$H_2 + Cl_2 \rightarrow 2HCl$$

Many organic materials (i.e., those based on carbon) will burn readily in halogen gases:

$$C_{10}H_{16} + 8Cl_2 \rightarrow 16HCl + 10C$$
turpentine

Nitrogen is not usually thought of as an oxidising agent or even a reactive element, but some metals will burn vigorously in this gas. Magnesium, aluminium and their alloys form nitrides in combustion reactions:

$$3Mg + N_2 \rightarrow Mg_3N_2$$
magnesium nitride

7.3 What makes a flame a flame?

> **There are two distinct types of flame: the premixed flame and the diffusion flame.**

If a pool of paraffin is heated, its temperature will rise and combustible vapours will evaporate from the surface. When the temperature of the liquid surface reaches about 50 – 55°C (the **firepoint**, see Section 7.6.3), the rate of evaporation is high enough for the vapours to be ignited by a small flame, or spark, and support continuous flaming above the surface. After the paraffin has been burning for some time, the surface of the fuel will be close to its boiling point, supplying flammable vapours to the flame.

Once a flame has been established and flammable vapours are rising from the fuel surface, heat and what are called **chain carriers** are produced where the flame reactions are occurring. A proportion of these will pass into the next layer of gas and start the oxidation and heat release processes there, rather as in a relay race. Chain carriers are believed to be atoms or fragments of molecules known as **free radicals** which are extremely reactive. The type of chemical reaction which occurs in the flame is known as **a chain reaction**.

In fact, there are two distinct types of flame: the premixed flame and the diffusion flame. They have different properties, though both are familiar from everyday experience. Understanding that each behaves differently is important: under different circumstances fuel and air can combine in different ways to produce very different results.

7.4 Laminar flow and turbulent flow

Before we discuss flames, it is useful to define two types of gas flow: laminar flow and turbulent flow.

Laminar flow [Figure 7.2] is steady flow in which two particles starting at any given point follow the same path. Particles never cross each other's paths, so the particle paths are bunched together like uncooked spaghetti in a packet. At any given time, the velocities of all particles on one path are the same, but the velocities of particles in different paths might be different. Laminar flow is associated with slow flow over smooth surfaces.

In **turbulent flow**, [Figure 7.2] there are random changes in velocity and direction of the flow, although the flow as a whole is moving in a definite direction. If we consider wind blowing down a street on a windy day, leaves and litter may be blown up, down, across and around, revealing local changes in the flow, but the general direction of the wind is still down the street. Turbulent flow tends to occur in fast flows over rough surfaces and around obstacles.

Figure 7.2 Laminar and turbulent flow.

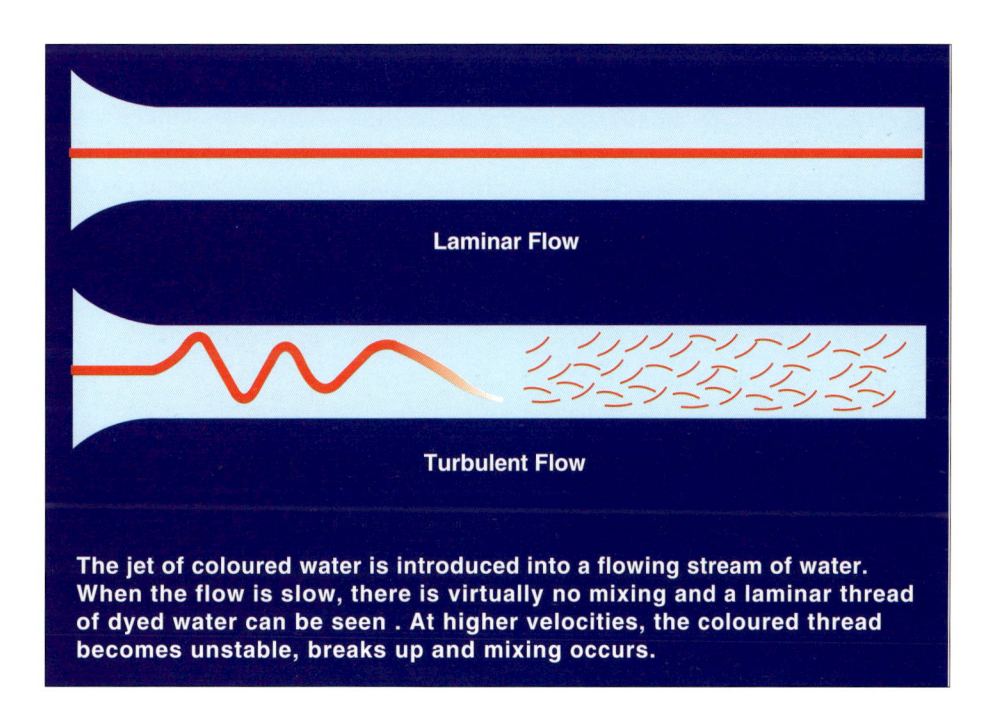

The jet of coloured water is introduced into a flowing stream of water. When the flow is slow, there is virtually no mixing and a laminar thread of dyed water can be seen. At higher velocities, the coloured thread becomes unstable, breaks up and mixing occurs.

Physics and Chemistry for Firefighters

7.5 Premixed and diffusion flames

As we have seen, a flame is the region in which the chemical reactions take place which turn unburnt fuel vapours into burnt gases – the combustion products: for example, methane and oxygen react to give carbon dioxide and water vapour.

A certain amount of heat energy is required to start this reaction but more heat is produced by the reaction than it takes to initially start it, so the burning process is self-sustaining.

Premixed flames occur when a fuel is well-mixed with an oxidant, e.g., 10% methane mixed with air. For ignition to occur, energy must be supplied to the system in the form of a spark or small flame. A self-sustaining flame will then be established around the ignition source and propagate outwards in all directions.

The flame consists of a zone where cold unburnt gas (reactants) is transformed into hot burnt gas (products). The flame zone of a premixed flame may be less than 1 mm thick. As the volume of the hot burnt gas is greater than that of the same mass of cold unburnt gas, the flame front is pushed outwards from the ignition point, like the skin of an inflating balloon.

Not every mixture of air and fuel will burn. Depending on the type of fuel and oxidant involved (air or pure oxygen, for example), a mixture initially at room temperature and pressure will only burn if the concentration of fuel lies between certain well-defined limits, called **flammability limits**. For example, mixtures of methane and air will only burn if the concentration of methane in air lies between 5% and 15%, whereas hydrogen will burn in air at concentrations between 4% and 75%.

The figures quoted for limits of flammability may vary as there are a number of factors which may slightly alter the value: pressure, temperature, dimensions of the test apparatus, direction of flame propagation and moisture content of the mixture all have some effect. (In general, the limits **widen** with rise in temperature.)

Within these ranges, there is an optimum mixture in which there is just sufficient fuel to use up all the oxygen. This is the **stoichiometric mixture**. Mixtures containing more fuel than the stoichiometric mixture are known as rich mixtures, and ones containing less fuel are lean mixtures. The stoichiometric mixture for methane is 9.4%. A 7% mixture is lean, while a 12% mixture is rich.

For each mixture of fuel and air between the flammability limits, there is a characteristic **burning velocity** at which a premixed flame will propagate through a stationary gas.

Burning velocities usually lie between 0.1 and 1.0 m/s. They tend to peak at the stoichiometric composition and fall away towards the flammability limits [Figure 7.3]. Burning velocity is dictated by the chemical processes involved – how quickly the fuel reacts with the oxygen. The methane and oxygen molecules do not simply combine instantaneously to form carbon dioxide and water vapour, but form free radicals and intermediates such as formaldehyde and carbon monoxide along the way to completing the reaction.

If the premixture flows into a flame with a laminar flow whose velocity is equal to the burning velocity of the mixture, the flame can be held stationary. This is how premixed flames on Bunsen burners, domestic gas rings etc., are held steady.

Local air currents and turbulence caused by obstacles can cause a flame to move at speeds much higher than the burning velocity. The speed at which a flame moves relative to an observer is the

Flammability Limits (% fuel/air by volume)

Gas	Lower limit	Upper limit
Hydrogen	4.0	75.0
Carbon monoxide	12.5	74.2
Methane	5.0	15.0
Butane	1.5	9.0
Ethylene	2.7	28.6
Acetylene	2.5	80.0

Figure 7.3 Flammability limits.

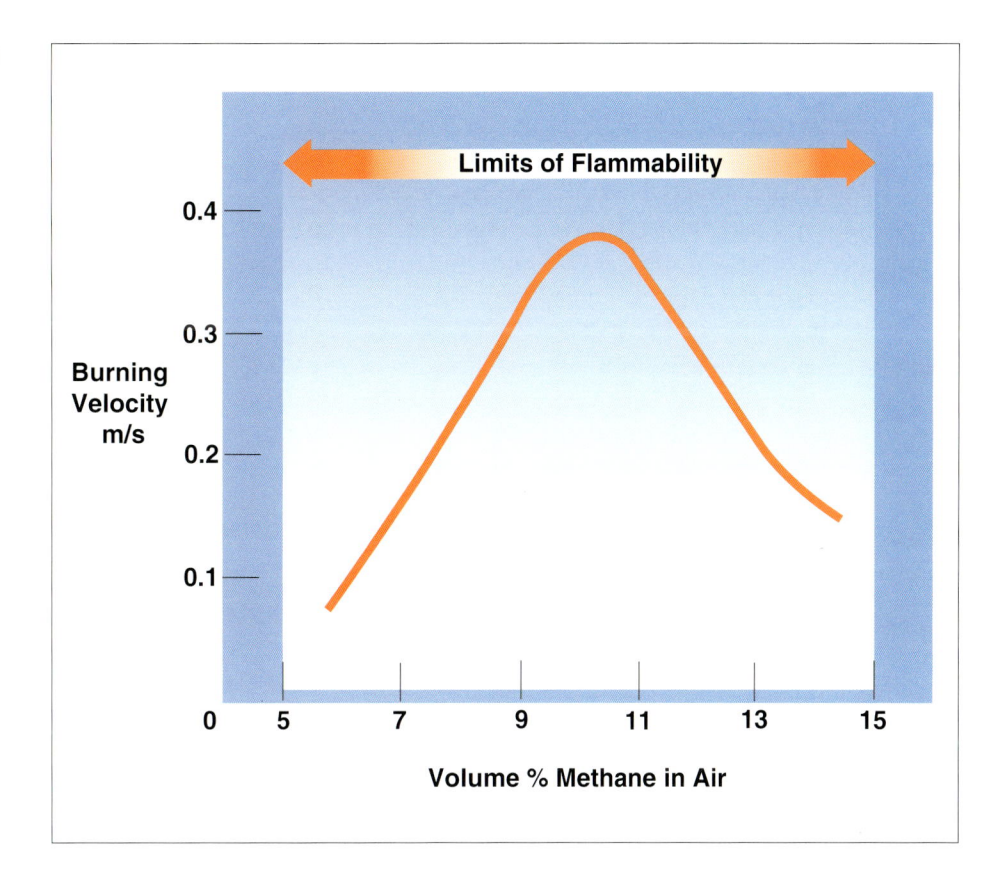

flame speed, which is different to the burning velocity. For example, the burning velocity of a stoichiometric methane-air flame is 0.45 m/s. If the unburnt gases are no longer stationary, the flame propagates at the local flow speed plus the burning velocity. As the flame gets faster, the flame front wrinkles as turbulence is produced in the unburnt gas, increasing the surface area of the flame front. This increases the reaction rate, increasing the rate at which burnt gas is produced, so pushing the flame front forward faster. In explosion situations, flame speeds of hundreds of metres per second can be achieved in gas-air mixtures, though the **burning velocity** of the mixture will be much lower than this. It is worthwhile making the distinction between these terms, as some older texts use flame speed to mean burning velocity. It is possible to achieve supersonic flame speeds, in which the combustion region is strongly coupled to a shock wave: this phenomenon is called **detonation**.

Diffusion flames occur at the interface where fuel vapour and air meet. Unlike premixed flames, the fuel vapour and the oxidant are separate prior to burning. The dominant process in the diffusion flame is the mixing process. The fuel vapour and oxygen mix with each other by molecular diffusion, which is a relatively slow process, though the high temperatures associated with flames increase the rate at which diffusion occurs.

Because diffusion flames exist only at the fuel-air interface, there is no equivalent of burning velocity, and no equivalent to rich or lean mixtures, or flammability limits.

Diffusion flames themselves fall into two broad types. In slow-burning diffusion flames, such as candle flames, the fuel vapour rises slowly from the wick in a laminar flow and molecular diffusion dominates. This type of flame is a **laminar diffusion flame**.

In industrial burners, fuel is injected at high velocity into the air as a spray or jet. Turbulence is induced at the interface where mixing takes place. This gives the flame an extremely large surface area in comparison to the relatively small surface area of the smooth fuel/air interface of the candle flame. In this turbulent case, it is the large interface area, rather than the rate of molecular diffusion, which determines the rate of mixing. This type of flame is a **turbulent diffusion flame**.

In a large fire (e.g., more than 1 m in diameter), the flames are turbulent diffusion flames, the turbulence generated by the strong buoyancy of the flames themselves. Inside the flame, there are regions of high temperature and low oxygen concentration where the fuel vapour is subjected to a mixture of pyrolysis (chemical decomposition in the absence of oxygen) and partial oxidation, leading to the formation of soot particles and products of incomplete combustion, in particular carbon monoxide (CO). These are the source of smoke, and of the gaseous species that render the fire products toxic.

7.6 Practical examples of premixed flames and diffusion flames

7.6.1 The Bunsen Burner

The Bunsen burner (Figure 7.4) should be familiar from school laboratories. It can produces both types of flame – premixed and diffusion. Gas is forced out of a jet at the base of the burner. If the air inlet collar at the bottom is open, air is entrained into the fuel flow and mixing occurs in the burner column. A pale blue conical flame is visible just above the top of the burner. This is a laminar premixed flame. When the air inlet to a Bunsen burner is closed, a yellow diffusion flame results.

7.6.2 A candle flame

When a match is held close to the wick of an unlit candle, the wax melts and rises up the wick by capillary action. There it evaporates, and a flame is established at the interface between the evaporating fuel and the surrounding air. The fuel and air are not mixed before burning, so this is a diffusion

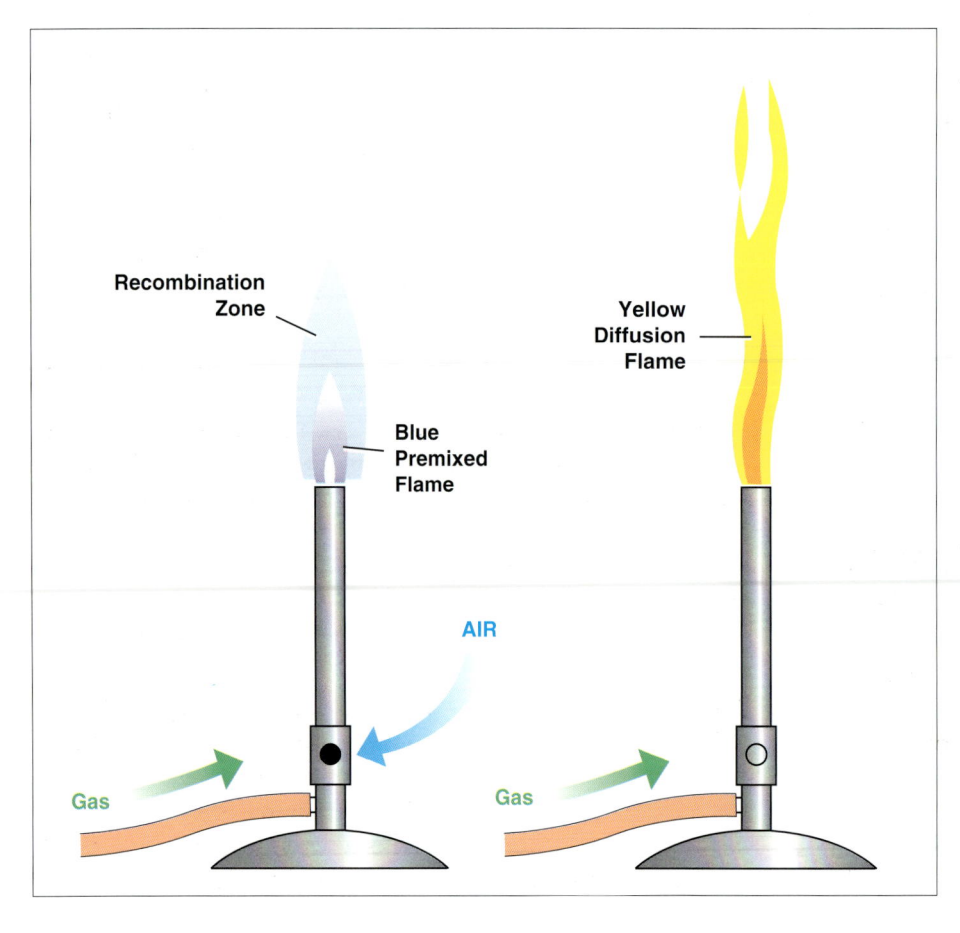

Figure 7.4 (a) Premixed flame on a Bunsen burner with full aeration; (b) diffusion flame.

flame. Once the flame is established, the process of melting, evaporation and burning is self-sustaining because heat is transferred from the flame back to the wick to sustain the melting and evaporation processes.

7.6.3 Flashpoint, firepoint and sustained fires

Imagine a dish of flammable liquid, such as paraffin. A region will exist above the liquid surface in which the evaporating fuel vapour is well mixed with air. If the paraffin is heated above about 40°C, this well-mixed region will become flammable – that is, the vapour concentration in air is above its lower flammability limit. **The lowest temperature at which this occurs is called the 'FLASHPOINT', the liquid temperature at which application of an ignition source will cause a flame to flash across the surface of the liquid.** This is a premixed flame moving through the vapour/air mixture but, just above the flashpoint, it burns out, or self-extinguishes, because it has consumed all the vapour. If heating is continued, a temperature will be reached at which ignition of the vapours will lead to a "flash", followed by the development of a sustained diffusion flame at the surface flame. **This temperature is known as the 'FIREPOINT', the lowest temperature at which the rate of supply of fuel vapours (by evaporation) can sustain the flame.**

There are several types of apparatus for determining the flash point of a liquid. The most common is the Pensky-Martens "closed cup" test in which the vapours cannot diffuse away from the surface, but can achieve a uniform concentration in the head space above the liquid surface. The Abel apparatus is also a closed cup test. The Cleveland "open cup" test gives a slightly higher flashpoint than the Pensky-Martens, but can be used to determine the firepoint. Clearly, it is always necessary to quote the method and the type of apparatus used, and whether the result is an "open cup" or "closed cup" flashpoint. Note that the flashpoint is affected slightly by pressure: values quoted in handbooks, etc., are adjusted to normal atmospheric pressure. Corrections should be considered for high altitude applications.

The term "flash fire" is used to describe what happens if the temperature of the fuel is much greater than the firepoint and a flammable vapour/air zone exists at some distance from the liquid surface. This may happen if there is a spillage of petrol (firepoint approx. -40°C) which forms a large pool. If an ignition source is introduced into the flammable zone, a premixed flame will flash back, igniting the fuel in the fuel-rich mixture above the liquid surface and giving rise to a large fire (turbulent diffusion flame).

In a sustained fire of this type, flames burn continuously above the surface until the fuel is consumed (or the fire is extinguished). In principle, combustible solids burn in the same way, although the formation of fuel vapours involves chemical decomposition of the solid which requires more energy than simple evaporation. For this reason, solids tend to burn much more slowly than combustible liquids.

7.6.4 Fireball

A fireball can occur when a mixture of vapour and mist droplets forms a cloud containing very little air, for example when a vessel containing pressurised liquid fuels, such as LPG, ruptures. The oxygen concentration within this cloud is far too low for premixed combustion to take place, but, if there is a source of ignition at the boundary between the fuel and the surrounding air, a premixed flame will flash through the flammable zone at the boundary, leading to the establishment of a diffusion flame. The fireball has then been established and will rise as it burns.

As burning progresses, instabilities are introduced in the surface of the flame increasing the surface area available for the reaction to take place. The fireball will increase in size until the fuel has been used up. Then it will shrink and extinguish.

7.6.5 Vapour cloud explosions

If a gas or vapour escapes under pressure from a rupture in a storage tank or pipeline, there is likely to be rapid mixing of the fuel with air, producing a cloud of fuel-air mixture, some of which will be of concentrations which fall between the flammability limits. If an ignition source is present in the cloud, a premixed flame will move outwards in all directions from the source. The flame will propagate through that region of the cloud which lies between the flammability limits.

After ignition, obstacles such as process vessels and pipework will introduce turbulence in the vapour/air mixture ahead of the flame, and cause wrinkling of the flame. This will increase the effective surface area of the flame, increasing the rate of burning, and the flame speed. This can generate overpressures which under severe conditions can cause blast damage. It is believed that this mechanism was responsible for the extensive damage caused at Flixborough in June 1974.

7.7 Ignition

For ignition to occur, sufficient heat energy is supplied to gaseous fuel and oxidant, either mixed to within flammability limits or at the interface between the fuel and the oxidant, to start a self-sustaining chemical reaction. This energy is generally supplied by a flame or a spark or a hot surface.

In this short section we will look at less obvious ways in which ignition can occur.

7.7.1 Spontaneous ignition temperature

This is the lowest temperature at which the substance will ignite spontaneously, that is the substance will burn without the application of a flame or other ignition source. This is sometimes referred to as the **auto-ignition temperature**.*

For some materials, the ignition temperature may be so low that there is a danger of them igniting under normal conditions, or in the range of temperatures that the material would experience during day-to-day use. Such materials are normally well documented and information available regarding their safe handling.

7.7.2 Self heating and spontaneous combustion

Certain materials may react with oxygen at room temperature. Compounds such as linseed oil which contain carbon-carbon double bands are very prone to this reaction, but any organic material stored in bulk quantities may be suspect, especially if it has been stored at an elevated temperature.

Imagine a pile of cloths soaked in linseed oil, which have been discarded after, for example a room has been decorated with paint or varnish containing a large proportion of linseed oil.

As the cloth is porous, oxygen in the air will be able to reach the centre of the pile. The linseed oil will oxidise slowly, even at room temperature, releasing heat. Because the centre of the pile is well insulated by the surrounding cloths, heat will build up and the temperature will rise. As the temperature rises, the reaction rate increases – roughly for every 10°C rise in temperature, the reaction rate doubles – so even more heat is given out and the temperature rises more quickly.

If heat is being produced more rapidly than it can escape, the temperature will continue to rise to a stage at which active combustion begins, usually well within the mass of material. Combustion will begin as a smouldering process, burning through to the surface of the pile where flaming combustion will be initiated.

Spontaneous combustion should be considered as a possible cause of a fire for which there is no obvious ignition source, but it is necessary to show that the material involved has the propensity to self heat, and that a sufficient quantity has been stored in such a way as to provide the necessary thermal insulation for the inside of the pile. A useful rule-of-thumb is that if the material does not produce a rigid char when it is heated, it is very unlikely to self-heat to ignition.

Sometimes the action of bacteria on certain organic materials can cause a rise in temperature eventually leading to active combustion (haystacks were prone to this). Powdered material, such as powdered coal used in power stations and some metals, can give rise to spontaneous combustion. Stocks of coal at power stations, if incorrectly stored in very large piles, can self-heat to ignition. In the manufacture of some plastics (e.g., polyurethane foams), the cross-linking process which creates the final molecular structure of the material is exothermic, and can lead to sponta-

---

* The auto-ignition temperature is not a true property of a material. It depends on how it is measured.

neous combustion if slabs of foam are stored before the process is complete (see Chapter 10).

7.7.3 Smouldering

Smouldering only occurs in porous materials which form a solid carbonaceous char when heated. Paper, sawdust, fibreboard and latex rubber can all undergo smouldering.

Smouldering is the combustion of a solid in an oxidising gas such as air, without the appearance of a flame. The process is very slow, but smouldering fires can go undiscovered for a very long time and can produce a large amount of smoke. This is flammable, but it must accumulate and reach its lower flammablity limit before it can be ignited. This has been known to happen on a few, fortunately rare, occasions (e.g., the Chatham Dockyard mattress fire in November 1974).

Smouldering will undergo a transition to flaming under favourable conditions. The best documented examples involve cigarette ignition of upholstered furniture. The mechanism is not fully understood, and it is impossible to predict how long after the commencement of smouldering that the transition will occur.

7.8 Hazards of oxidising agents

Nearly all combustion reactions involve oxidation which in its most simple form is combination with oxygen, such as the combustion of hydrogen:

$$2H_2 + O_2 = 2H_2O$$

The oxygen in this case may be called an oxidising agent. The word oxidation also has a broader meaning where elements other than oxygen may be considered as oxidising agents. For example, most metals will react with chlorine, and other halogens: this is also a type of oxidation.

$$Mg + Cl_2 = MgCl_2$$

Here chlorine is the oxidising agent.

There are certain compounds which do not necessarily burn themselves but, on decomposition, release oxygen which can greatly assist a combustion reaction. Some of these compounds may be relatively stable at room temperature but at high temperatures they could be extremely hazardous.

Some of the more common oxidising agents are considered below.

7.8.1 Nitric acid and inorganic nitrates

Concentrated nitric acid is a very powerful oxidising agent and reacts vigorously with many organic compounds. Carbon itself reacts with the hot acid, in the following way:

$$C + 4HNO_3 = CO_2 + 4NO_2 + 2H_2O$$

When the concentrated acid mixes with carbonaceous (carbon-containing) material there is a violent reaction giving off a great deal of heat and nitrogen dioxide (nitrous fumes). Clearly, carbonaceous materials like sawdust and wood chippings must never be used to soak up a spillage of this acid. The nitrates (salts of nitric acid) are also good oxidising agents. They are used in large quantities in industry and agriculture. An example is the use of molten nitrate salt baths for the heat treatment of metals.

If they are strongly heated the nitrates of sodium and potassium give off oxygen and the metal nitrite:

$$2KNO_3 = 2KNO_2 + O_2$$

Most other metal nitrates form the metal oxide, giving off nitrogen dioxide ('nitrous fumes') and oxygen.

Ammonium nitrate is widely used as an agricultural fertiliser under various trade names. It is a white crystalline solid, very soluble in water (all nitrates are soluble in water). It does not burn by itself, but mixing it with a fuel (e.g., sugar) produces a powerful explosive. It decomposes violently when heated giving nitrous oxide and water:

$$NH_4NO_3 = N_2O + 2H_2O$$

Brown nitrous fumes (NO_2) are also given off on heating; decomposition is complex. These fumes of nitrogen dioxide will support combustion in a similar manner to oxygen.

Ammonium nitrate can detonate, but only large quantities, and under extreme conditions.

7.8.2 Permanganates

Of the permanganates, sodium ($NaMnO_4$) and potassium ($KMnO_4$) permanganates are the most common. They are powerful oxidising agents and may react violently with oxidisable organic materials. When mixed with glycerol (glycerine), spontaneous ignition occurs. With concentrated hydrochloric acid, permanganates produce the highly toxic chlorine gas as a result of oxidation.

7.8.3 Chlorates

Chlorates are often used as their sodium or potassium compounds. On heating, oxygen is released:

$$2KClO_3 \rightarrow 2KCl + 3O_2$$

Very violent reactions occur on contact with oxidisable materials and may occur merely by friction.

Potassium perchlorate ($KClO_4$) has a similar chemical formula, but is, in fact, stable. Anhydrous perchloric acid ($HClO_4$) is a powerful oxidising agent and will explode on heating. Sodium chlorate is used as a weed killer and has also been used in home-made explosives.

7.8.4 Chromates and dichromates

The most common compounds of this type are potassium chromate (K_2CrO_4) and potassium dichromate ($K_2Cr_2O_7$); these materials are yellow and orange respectively and are oxidising agents. They are soluble in water and will produce a highly combustible mixture with oxidisable substances.

7.8.5 Inorganic peroxides

Peroxides are a group of compounds which contain a higher proportion of oxygen than the 'normal' oxide. This extra oxygen is easily liberated, making these compounds good oxidising agents. Inorganic peroxides may be considered to derive from hydrogen peroxide (H_2O_2). Pure hydrogen peroxide is a clear viscous liquid with a specific gravity of 1.46 (at 0°C). It is soluble in water and is used at various concentrations. Above 70 per cent concentration in water it is a powerful oxidising agent and decomposes explosively:

$$2H_2O_2 \rightarrow O_2 + 2H_2O$$

This decomposition may occur on heating, but can also occur in by the presence of a catalyst: small traces of metallic dust, charcoal or even strong light may be sufficient. Concentrated solutions of hydrogen peroxide are often known as 'high test peroxide' (HTP).

Common metal peroxides, derived from hydrogen peroxide, are those of sodium (Na_2O_2) and barium (BaO_2). Sodium peroxide is a pale yellow solid which reacts vigorously with water, releasing oxygen:

$$2Na_2O_2 + 2H_2O \rightarrow O_2 + 4NaOH$$

A great deal of heat is released in this reaction and this could cause a fire in any nearby combustible material. The fire would be made worse by the oxygen evolved.

Sodium peroxide can absorb carbon dioxide, releasing oxygen as a product:

$$2Na_2O_2 + 2CO_2 \rightarrow 2Na_2CO_3 + O_2$$

7.8.6 Organic oxidising agents

When nitric acid reacts with organic compounds, two important types of substance are formed: organic nitrates ($-NO_3$) and nitro-compounds ($-NO_2$).

These compounds are oxidising agents and furthermore they carry oxidisable carbon-containing material within their own molecules. Consequently, both the organic nitrates and the nitro-compounds are highly flammable. Some that contain several nitrate or nitro groups in the molecule are explosive. Typical examples are glyceryl trinitrate (used in dynamite) and trinitrotoluene (TNT) – an important military explosive.

Most organic nitrates and nitro- compounds are toxic and many of them, including glyceryl trinitrate, may be absorbed through the skin.

7.8.7 Organic peroxides and hydroperoxides

These compounds have a structure similar to that of hydrogen peroxide (H_2O_2, arranged as H-O-O-H), with both hydrogen atoms replaced by organic groups, thus forming an organic peroxide. If only one hydrogen is replaced, a hydroperoxide is formed.

As would be expected peroxides and hydroperoxides are powerful oxidising agents and, because there is a carbon-containing part of the molecule which can be oxidised, they are highly flammable. Some are explosive and sensitive to heat and mechanical shock. Because of this they are often diluted or 'damped down' with either water or stable esters.

Peroxides are extensively used as catalysts, especially in the plastics industry. They are toxic and are especially irritating to the skin, eyes and mucous membranes. Skin contact and breathing of vapours should be avoided. In all respects, organic peroxides and hydroperoxides should be treated with extreme caution.

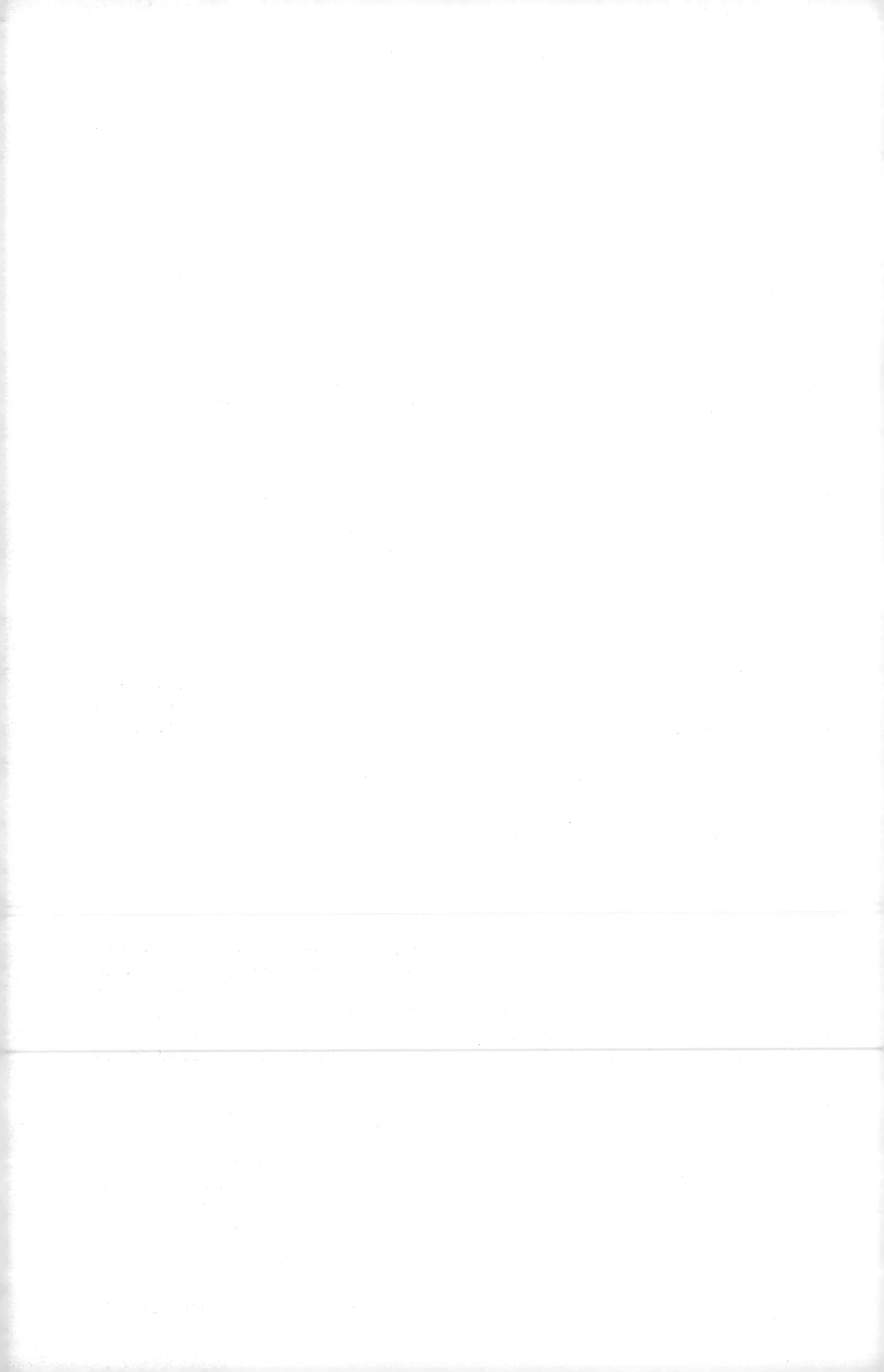

Physics and Chemistry for Firefighters

Chapter 8

Chapter 8 – Simple organic substances

Carbon forms a very large number of compounds, especially with hydrogen, oxygen, nitrogen and the halogens. It forms so many compounds that chemistry is divided into two branches:

- organic chemistry which deals with the chemistry of the carbon compounds; and

- inorganic chemistry which deals with the chemistry of all the other elements.

There are believed to be over a million stable carbon compounds, which explains why a separate branch of chemistry is necessary to study them.

Carbon atoms differ from almost every other type of atom in that they are able to link up with other carbon atoms and form chains and rings. Most other atoms only join with others of the same kind in twos or threes. In all these organic compounds the valency of carbon is always four.

Organic chemicals are divided into two classes:

- aliphatic compounds, which contain chains of carbon atoms; and

- aromatic compounds which contain a special kind of ring of six carbon atoms, known as a benzene ring.

Most organic chemicals are capable of burning. Our most important fuels, such as natural gas, petrol, paraffin and diesel oil are mixtures of organic compounds which contain carbon and hydrogen – the hydrocarbons.

8.1 Aliphatic hydrocarbons (paraffins or alkanes)

The aliphatic hydrocarbons are a series of compounds containing only carbon and hydrogen. The simplest member of paraffins, or alkanes, is methane, the main constituent of natural gas. It has the formula CH_4 and the structure of the molecule is conveniently represented as:

```
      H
      |
H —— C —— H
      |
      H
```
Methane, CH_4

The carbon atom uses each of its four valencies to join it to four hydrogen atoms which each have a valency of one. The CH_4 molecule can also be regarded as a combination of the group:

```
      H
      |
H —— C ——      called a methyl group
      |
      H
```

with a hydrogen atom. Methane has well-defined chemical and physical properties. It is a relatively unreactive gas, although it is flammable and forms explosive mixtures with air at concentrations between 5% and 15% by volume. In common with other hydrocarbons, it burns completely to produce carbon dioxide and water:

$$CH_4 + 2O_2 = CO_2 + 2H_2O.$$

Larger molecules are built up by linking the carbon atoms together in chains. Hydrogen atoms are attached to the carbon atoms in accordance with the valency rules. For example:

Physics and Chemistry for Firefighters **59**

Methane, CH₄ with an additional C atom becomes Ethane, C₂H₆

Ethane (C₂H₆) is another constituent of natural gas. Because the molecule is larger, the physical properties are different from methane. The boiling point, melting point and vapour density of ethane are higher than those of methane, whereas its spontaneous ignition temperature is lower.

Another increment in the chain length of the molecule results in propane (C₃H₈), a constituent of liquefied petroleum gas.

Ethane, C₂H₆ with an additional C atom becomes Propane, C₃H₈

Propane is chemically similar to ethane and methane, but once again, the physical properties differ (see Table 8.1).

In principle, the carbon chain can be extended indefinitely, until the chain consists of many thousands of carbon atoms, as in polyethylene (see Chapter 9). The longer the chain, the higher the boiling and melting points of the substance. Methane, ethane, propane and butane are all gases at room temperature and pressure, but heavier molecules, starting at pentane, C₅H₁₂, are liquids, and from hexadecane, C₁₆H₃₄ they are solids. Compounds near C₈H₁₈ (octane) are found in petrol, those near C₁₀H₂₂, in paraffin, those near C₁₄H₃₀ in diesel oil, those near C₁₈H₃₈ in petroleum jelly (Vaseline) and those near C₂₅H₅₂ in paraffin wax. The following points may be noted concerning these compounds:

- they form a series in which each differs from the next by one -CH₂ unit;

- they have similar chemical properties; and

- their physical properties vary in a regular way.

As the number of carbon atoms increases:

- melting point, vapour density, boiling point and flash point **increase**; and

- solubility in water and spontaneous ignition temperature **decrease**.

8.2 Unsaturated aliphatic hydrocarbons

8.2.1 Olefines or alkenes

There is another series of aliphatic compounds known as the **olefins or alkenes**. The first member of the series is ethylene (C₂H₄), the formula for which is represented as:

$$H_2C=CH_2$$

where there is a double bond between the two carbon atoms. The carbon still has its valency of 4 and hydrogen that of 1, but each carbon atom uses two of its valency bonds to link to the other carbon atom. Compounds containing double bonds (and those with triple bonds) are known as **unsaturated** compounds.

Unsaturated compounds are much more reactive than the paraffins. Not only do they burn, but they react readily with chlorine, hydrogen chloride, bromine, and other reagents. For example:

$$H_2C=CH_2 \;(\text{Ethylene}) + Cl_2 \longrightarrow H_2ClC-CClH_2 \;(\text{Ethylene dichloride or dichloroethane})$$

The reactivity of ethylene makes it an important starting material in the production of plastics and other synthetic materials (see Chapter 9).

Fire Service Manual

Table 8.1 *The Paraffins or Alkanes*

	Name	Formula	Vapour density (A = 1)	Melting point (°C)	Boiling point (°C)	Flash point (°C)	Flammable limits (% in air)	Self-ignition temperature (°C)
Gases	Methane	CH_4	0.554	−183	−161	Gas	5 to 15	538
	Ethane	C_2H_6	1.04	−172	−89	Gas	3.3 to 12.5	510
	Propane	C_3H_8	1.52	−187	−42	−104	2.4 to 9.5	510
	Isomers n-Butane	C_4H_{10}	2.046	−138.6	−0.6	−60	1.5 to 9.0	466
	iso-Butane	C_4H_{10}	2.046	−160	−12	Gas	1.8 to 8.4	545
Liquids	n-Pentane	C_5H_{12}	2.48	−130	36	<−40	1.4 to 7.8	309
	n-Hexane	C_6H_{14}	2.97	−95.6	69	−23	1.2 to 7.4	260
	↓ n-Hexadecane	$C_{16}H_{34}$	7.8	18	287	>100	–	205
Solids	n-Heptadecane	$C_{17}H_{36}$	8.33	22	303	148	–	200
	n-Octadecane	$C_{18}H_{38}$	8.81	28	308	165	–	200

Other olefins can be obtained by progressively increasing the length of the carbon chain. As with the paraffins, the physical properties alter in a regular way as the size of the molecule increases (see Table 8.2).

8.2.2 Acetylenes, or alkynes

Acetylenes contain a carbon-carbon triple bond. The only important member is the gas acetylene (C_2H_2). It is an unsaturated compound and the only way of arranging the normal valency bonds of the carbon and hydrogen is:

$$H-C\equiv C-H$$
Acetylene

Here, each carbon atom uses three of the available valencies to make three bonds with the other. This triple bond makes acetylene very reactive: it can explode on exposure to heat or mechanical shock, even when air or oxygen are absent. Acetylene is

Physics and Chemistry for Firefighters 61

Table 8.2 The Olefines or Alkenes

Name	Formula	Structure	Vapour density (A = 1)	Melting point (°C)	Boiling point (°C)	Flash point (°C)	Flammable limits (% in air)	Self-ignition temperature (°C)
Ethylene	C_2H_4	H₂C=CH₂	0.98	−169	−103.9	Gas	2.7 to 28.6	450
Propylene	C_3H_6	H−CH=CH−CH₃	1.5	−185	−48	Gas	2 to 11	495
The Butylenes	C_4H_8	(various isomers)	1.93	about −185	about −6.3	<−80	1.7 to 10.0	384
Gases								

* The olefines are liquids from C_5H_{10} to $C_{16}H_{32}$, and then solids.

Table 8.3 The Acetylenes or Alkynes

Name	Formula	Structure	Vapour density (A = 1)	Melting point (°C)	Boiling point (°C)	Flash point (°C)	Flammable limits (% in air)	Self-ignition temperature (°C)
Acetylene	C_2H_2	H−C≡C−H	0.91	−81	−84†	−17.7	2.5 to 80	335
Methylacetylene	C_3H_4	H−C≡C−CH₃	1.38	−102.7	−23	Gas	1.7 to 11.7	−
*Gases								

* The acetylenes are liquids from C_4H_6 to $C_{17}H_{32}$, and then solids.
† Sublimes

flammable and forms mixtures in air with wide flammability limits (2.5% – 80%). Some of its physical properties are given in Table 8.3. Acetylene is used in the manufacture of plastics (e.g., PVC), certain chemicals and in oxyacetylene welding. It is stored by dissolving it in acetone, which is absorbed on an inert porous material contained in cylinders.

8.3 Aromatic hydrocarbons

The simplest member of the aromatic hydrocarbons is benzene, C_6H_6. It has a unique structure, consisting of six carbon atoms arranged in a ring, apparently with alternating single and double bonds:

However, the six carbon atoms are linked in a special way, and benzene does not behave like an olefin. In fact, the benzene ring is remarkably stable, so that aromatic compounds – those containing the benzene ring, such as toluene – are less reactive than olefins.

Benzene is a flammable liquid, but when it burns, the aromatic ring proves difficult to oxidise because of its stability: the high proportion of carbon in the molecule leads to the formation of a thick black smoke. It is no coincidence that the precursors of smoke in flames from any fuel have been found to have an aromatic structure. These are formed in the flame in regions where temperatures are high, but the oxygen concentration is low. If the fuel already contains an aromatic structure, it can be said quite categorically that it will burn with a very smoky flame.

Other aromatic compounds are formed by replacing the hydrogen atoms of benzene by other atoms or groups of atoms, such as methyl radicals. For example:

Again, the physical properties of the members of this series vary in a regular way as the molecular weight increases, and the chemical properties remain similar. Some aromatic compounds, especially toluene and the xylenes are important solvents.

It is worth noting that hydrocarbons do not dissolve to any extent in water, but will float as their specific gravity is less than 1. Some aromatic compounds are toxic; benzene for example is highly toxic, both as a vapour and by skin absorption of the liquid. Many, more complex, aromatic compounds have been identified as carcinogens. Similar compounds are to be found in smoke.

8.4 Liquefied petroleum gases (LPG)

Propane (C_3H_8) and butane (C_4H_{10}) are gases at room temperature and pressure, but are easily liquefied using pressure alone. A very small amount of liquid will produce a great volume of gas and so by liquefying the gas a large amount can be stored in a small volume. As both gases are highly flammable and are widely used as fuel gases, installations containing the liquid gases are very widespread.

The important property for LPG is the critical temperature – the temperature above which it is impossible to liquefy a gas by pressure alone (see Section 4.4). For propane, the critical temperature is 96.7°C and for butane, 152°C. When in the right kind of container they will evaporate, increasing the pressure until there is no further net evaporation. Thus, inside each liquid gas container there is a liquid with pressurised vapour above it. As the gas is let out for use as fuel, more liquid evaporates to keep the pressure of the vapour at its original value.

As discussed, both propane and butane are highly flammable. If propane liquid escapes, it will quickly boil into a large amount of flammable vapour: 1 litre of liquid will produce 270 litres of vapour.

The vapours of both propane and butane are heavier than air and will seek lower ground; they are odourless and colourless, though very frequently, a stenching agent (mercaptan) is added.

Physics and Chemistry for Firefighters 63

When propane and butane evaporate they take heat from their surroundings. Propane has a boiling point of -42°C at atmospheric pressure, so at normal ambient temperatures and pressures, the liquid boils easily. This applies to vast majority of practical situations. However, butane has a boiling point of around -1°C, so that in winter conditions, the vapour pressure may be too low to provide a flow of fuel vapour. For this reason, "LPG" generally consists of a mixture of propane and butane.

> **In dealing with LPG it is vital to realise that above the critical temperature the substances can only exist as gases. Cylinders heated above this temperature are, therefore, likely to explode.**

Full details of the storage and fire-fighting techniques associated with liquefied petroleum gases will be found in the *Manual of Firemanship, Part 6c*.

8.5 Simple oxygen-containing compounds derived from hydrocarbons

There are many different types of organic compound which contain oxygen in addition to carbon and hydrogen. Some are reactive, and may be encountered in industrial processes (e.g., aldehydes) whereas others are relatively unreactive and are used as solvents (e.g., ketones). The more important types are discussed briefly.

8.5.1 Alcohols

The structure of the commonest alcohols is similar to that of paraffins, but with one of the hydrogen atoms replaced by a hydroxyl group O—H:

Methane, CH_4 — with one H atom replaced by an O—H, group becomes — Methyl alcohol CH_3OH (Methanol)

The whole series of alcohols is formed by adding -CH_2 groups to methanol:

- ethyl alcohol or ethanol (C_2H_5OH);
- n-propyl alcohol or n-propanol (C_3H_7OH); and
- n-butyl alcohol or n-butanol (C_4H_9OH).

As the molecular weight increases, there is a **general increase** in the:

- melting point;
- boiling point; and
- flash point.

This is accompanied by a **decrease** in the:

- solubility in water; and
- spontaneous ignition temperature.

The first few members of the series dissolve completely in water but members higher than butyl alcohol are only slightly soluble. All alcohols are less dense than water: the insoluble ones float.

Chemically the alcohols resemble each other. The first members of the series are highly flammable liquids. Ethyl alcohol is sometimes used as a fuel and is also used in rocket propulsion systems. Alcohols are also intermediaries in various chemical processes. Methanol and ethanol are widely used as solvents in industry and ethanol is the most important ingredient of beer, wines and spirits. Propanol and butanol are also used as solvents and some of the higher alcohols are used to make detergents. Nearly all alcohols are to a greater or lesser extent toxic according to type.

8.5.2 Aldehydes

These compounds all contain the group:

$$\begin{array}{c} H \\ | \\ -C \\ \| \\ O \end{array}$$

attached to such organic groups as methyl (CH$_3$-), ethyl (C$_2$H$_5$-) and so on. The simplest member of the group is formaldehyde:

$$\begin{array}{c} H \\ \diagdown \\ C=O \\ \diagup \\ H \end{array}$$

with the formula CH$_2$O. Formaldehyde is a colourless, flammable gas with a pungent suffocating smell, though it is more usual to find it as a 4% solution in water, which also contains a little methanol. This solution is called formalin and gives off flammable vapour if heated above its flash point. This varies according to the formaldehyde and methanol concentration. The vapour is toxic. Formaldehyde is used in the manufacture of several plastics, as an antiseptic and as a preservative of anatomical specimens.

The next member of the series, acetaldehyde (CH$_3$CHO) has the structure:

$$H-\overset{\overset{\displaystyle H}{|}}{\underset{\underset{\displaystyle H}{|}}{C}}-\overset{\overset{\displaystyle H}{|}}{C}=O$$

and differs from formaldehyde by one CH$_2$ group. It is a colourless liquid with a strong fruity smell. The compound dissolves readily in water, is flammable and toxic and the vapour forms explosive mixtures with air. It is used as an intermediate in the manufacture of other chemicals and plastics.

The higher aldehydes (i.e. containing more than two carbon atoms) are rather less important. The physical properties of these materials vary in the usual way as the molecular weight increases.

8.5.3 Ketones

The simplest ketone is acetone [(CH$_3$)$_2$CO] which has the structure:

$$H-\overset{\overset{\displaystyle H}{|}}{\underset{\underset{\displaystyle H}{|}}{C}}-\overset{}{\underset{\underset{\displaystyle O}{\|}}{C}}-\overset{\overset{\displaystyle H}{|}}{\underset{\underset{\displaystyle H}{|}}{C}}-H$$

Acetone is by far the most commercially important member of the series. It is a colourless, highly flammable liquid which is readily soluble in water and has a minty smell. It is toxic to the extent that high concentrations have an anaesthetic effect and the liquid dissolves the fats out of the skin, and so can give rise to dermatitis and skin irritations. Acetone is a very important industrial solvent for materials such as paint removers, cellulose acetate, fats, waxes and acetylene (see Section 8.2).

The next member of the series is methyl ethyl ketone (MEK):

$$\begin{array}{c} C_2H_5 \\ \diagdown \\ C=O \\ \diagup \\ CH_3 \end{array}$$

It is another important industrial solvent and closely resembles acetone, but with a higher flashpoint, etc.

8.5.4 Carboxylic acids

The carboxylic acids or 'fatty acids' are a group of weak acids, related to the aliphatic hydrocarbons with a similar chain structure. They all contain the group:

Physics and Chemistry for Firefighters 65

$$\mathrm{-C} \underset{O-H}{\overset{\displaystyle O}{\diagdown}}$$

which may be found attached to various organic groups such as methyl (-CH$_3$) and ethyl (-C$_2$H$_6$). The first member of the carboxylic acid series is formic acid, HCOOH:

$$\mathrm{H-C} \underset{O-H}{\overset{\displaystyle O}{\diagdown}}$$

Formic acid is a colourless liquid with a pungent smell. It is toxic and can cause burns on the skin. It is used in the textile industry, in electroplating and in the leather and rubber industries.

The next member of the series is acetic acid (CH$_3$COOH), which is present as a dilute solution in vinegar. It is flammable, dissolves readily in water and can burn the skin if concentrated. The vapour and the concentrated acid are toxic. It is used as a solvent in chemical manufacture.

The properties of some of these acids are given in Table 8.4.

8.5.5 Esters

The esters can be thought of as being derived from the carboxylic acids by the replacement of the hydrogen atom in the COOH group by a methyl or other radical. For example, "ethyl acetate" (see opposite).

Esters are flammable, colourless liquids or solids and are usually only slightly soluble in water, on which they float. They have fruity smells and are often found in fruit and in scents. The higher solid members of the series are found in beeswax. The esters are used as solvents, and in pharmaceuticals, perfumery and foodstuffs. Some properties of a few esters are given in Table 8.5.

8.5.6 Ethers

These all contain an oxygen atom -O- which joins two organic groups such as methyl or ethyl groups. The only one of commercial significance is diethyl ether (C$_2$H$_5$-O-C$_2$H$_5$). This substance is often referred to merely as 'ether'. It is a colourless, highly flammable, volatile liquid with a characteristic smell. It is less dense than water and immiscible in it. It is toxic in high concentrations and at lower concentrations has an anaesthetic effect.

Table 8.4 *Carboxylic acids*

Name	Formula	Structure	Vapour density (A = 1)	Melting point (°C)	Boiling point (°C)	Flash point (°C)	Flammable limits (% in air)	Self-ignition temperature (°C)
Formic acid	HCOOH		1.59	8	101	69	18 to 51	600
Acetic acid	CH$_3$COOH		2.07	16.6	118	45	5.4 to 16	566
Propionic acid	C$_2$H$_5$COOH		2.56	−22	141	54	2.9 to 14.8	577
n-butyric acid	C$_3$H$_7$COOH		3.04	−7.9	163.5	72	2 to 10	452

Acetic acid, CH₃COOH — with the *H replaced by the ethyl radical C₂H₅, becomes — Ethyl acetate, CH₃COOC₂H₅

Table 8.5 *Esters*

Name	Formula	Structure	Vapour density (A = 1)	Melting point (°C)	Boiling point (°C)	Flash point (°C)	Flammable limits (% in air)	Self-ignition temperature (°C)
Ethyl formate	HCOOC₂H₅		2.55	-79	-54	-20	28 to 16.5	455
Ethyl acetate	CH₂COOC₂H₅		3.04	-84	77	-4	2 to 11.5	>427
Ethyl butyrate	C₃H₇COOC₂H₅		4	−93	121	25	–	463
Amyl acetate	CH₃COOC₅H₁₁		4.5	−78.5	148	25	1.1 to 7.5	399

Diethyl ether has a boiling point of 34.5°C, a flash point of -4.8°C, flammable limits of 1.85 to 36.5% in air and a self-ignition temperature of 180°C. It may contain a substance known as ether peroxide which, if the ether is evaporated to dryness, can cause an explosion.

Physics and Chemistry for Firefighters

Chapter 9

Chapter 9 – Polymers

9.1 Polymers

Many organic solids, including wood, plastics and rubbers, are polymers. This means that the molecules of which they are composed consist of very long chains of carbon atoms which can consist of many thousands of atoms.

For many years now, chemists have been able to create or synthesise polymer molecules in the laboratory. Many of these have passed into commercial use as plastics and synthetic rubbers. Polymers are formed by taking small molecules with two or more reactive groups and arranging for these to link up end to end and form long chains. For example in ethylene $H_2C=CH_2$, the double bond can be "opened" or broken to give:

H_2C——CH_2

which will very rapidly combine with other molecules of the same type to produce polyethylene:

-CH_2-CH_2-CH_2-CH_2-CH_2-CH_2-

which consists of a long chain of -CH_2- groups joined to each other. In a case such as this we call ethylene the **monomer**, polyethylene the resulting **polymer** and the process **polymerisation**.

Simple straight-chain polymers of this type are well known; many of them have small side groups attached to the chain, as in polypropylene:

or polystyrene:

Here the symbol:

stands for the benzene ring, which is the simplest aromatic structure (Section 8.3).

Polymers such as polyethylene, polypropylene and polystyrene soften and eventually melt at temperatures in excess of 100-150°C. Such materials are called **thermoplastics**.

Thermosetting plastics do not melt but break down and char on heating. In these plastics, the long chains are also linked together sideways by carbon-carbon bonds: the material is said to be **cross-linked**. Figure 9.1 illustrates this point, the lines representing polymer molecules and their cross-links.

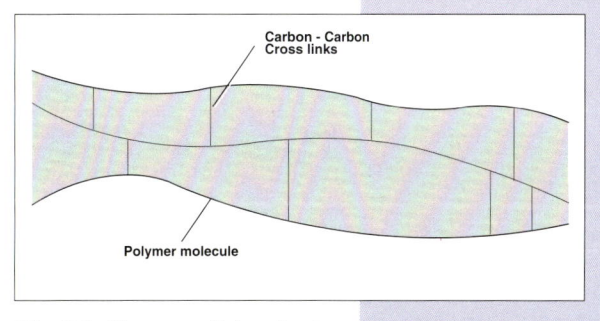

Fig. 9.1 The cross-links of polymer molecules of thermosetting plastics.

Industry uses the fact that thermoplastics soften or melt when they are processed. The same techniques can not be used in thermosetting plastics. These have to be processed as short chain molecules and then heated or a catalyst added to make the molecules cross-link.

Plastics often have other materials mixed into them to improve their properties or to make them cheaper:

- **Inert fillers** such as china clay, wood flour and carbon black, or, in laminated plastics, sheets of paper or glass-cloth;

- **Fire retardants**, to make the polymer more difficult to ignite;

- **Plasticisers**, mixed with some thermoplastics to make them more pliable (e.g., PVC insulation);

- **Stabilisers** to inhibit degradation due to atmospheric oxidation, attack by sunlight or decomposition under conditions of mild heating; and

- **Colouring materials**.

9.2 Fire hazards

Like any combustible material in a fire situation, plastics and rubbers may:

- **give off toxic and corrosive gases; and**

- **give off large quantities of smoke, often in a very short space of time.**

However, many synthetic polymers produce much more smoke than "traditional" materials such as wood, and the rate of fire growth may be much greater, particularly if the material melts and drips, spreading fire as a burning liquid.

9.2.1 Toxic and corrosive gases

If plastics only contain the elements carbon and hydrogen, or carbon, hydrogen and oxygen, the main toxic gas to be expected is carbon monoxide (CO) which is formed when all organic materials are burned in quantity. The amount of CO produced increases if there is a relative shortage of oxygen. This gas is a well-known hazard to fire fighters. As it is odourless, colourless and produced in large quantities in fires in buildings, it is responsible for most of the fire fatalities. It can also cause the deaths of people who are confined to unventilated rooms with faulty gas heaters.

Many other toxic products are produced from plastics containing only carbon, hydrogen and oxygen. A whole range of toxic and corrosive species are produced under conditions of poor ventilation. Their nature depends on the structure of the polymer, and can include aldehydes and many other partially oxidised products.

Many plastics contain nitrogen in addition to carbon, hydrogen and oxygen. Plastics in this category include cellulose nitrate, nylon, polyurethane foams, melamine-formaldehyde plastics, urea-formaldehyde plastics, ABS (acrylonitrile-butadiene-styrene), some epoxy resins and nitrile rubbers. (Note that certain "natural polymers" such as wool and silk also contain nitrogen.) The fire products from these will contain nitrogen-containing species such as organic nitriles, hydrogen cyanide and NO_2. All of these are toxic. Intense, well ventilated burning will convert most of the original nitrogen into NO_2.

The class of material known as polyurethane foams (PUF) includes the "standard PUF" introduced in the 1970's as well as the newer "combustion modified foams" which began to appear in the 1980's. The standard foams have been extensively investigated and are known to produce appreciable amounts of CO and HCN in the fire gases. Other, highly toxic nitrogen-containing products may be produced, depending on the type of polyurethane foam and the conditions of burning.

Chlorine is present in polyvinyl chloride (PVC) and certain related co-polymers, in neoprene and in certain types of self-extinguishing fibre-glass polyester resin. In PVC fires, almost all the chlorine goes to form hydrogen chloride (HCl) gas in the fire gases. HCl is both toxic and corrosive, having a very sharp smell and forming a corrosive solution with water (hydrochloric acid). Apart from corroding many metals, the acid may cause long term changes in alkaline mortar. Ferroconcrete may be much less affected; nevertheless, copious washing after incidents involving PVC is desirable. Other chlorine containing polymers may also give hydrochloric acid gas and possibly other chlorine containing toxic compounds as well.

PTFE (poly-tetrafluoroethylene – 'Teflon') and some related materials, such as 'Kel-F' and some synthetic rubbers, sometimes known as 'vitons', contain fluorine.

If these materials are overheated, toxic fluorinated gases are produced. If these are inhaled through a lighted cigarette they become more dangerous. Toxic products from decomposing fluorinated materials should not be inhaled.

9.2.2 Smoke

Fires involving materials which contain the aromatic (benzene) ring structure (Section 8.3) will tend to produce large quantities of smoke. These include polyurethanes, phenol-formaldehyde resins, polystyrene (Section 9.1), polyesters, epoxy resins and polycarbonates.

If materials of this type ignite readily and burn rapidly in a fire (e.g., the early polyurethane foams), very large quantities of thick black smoke will be evolved in a very short period of time, and can lead to rapid smoke-logging of escape routes, etc.

Although PVC does not contain the aromatic structure, it produces large amounts of smoke in fires, because it decomposes in the solid phase to produce aromatic structures.

9.2.3 Burning tars or droplets

Thermoplastics melt on heating and so in a fire may form burning droplets which could help the fire to spread. Although polyurethane foams are not technically thermoplastics, they do give burning drops of tar in a fire. On the other hand PVC, which is a thermoplastic, does not give burning droplets, but merely forms a tarry coke-like product.

9.2.4 Exotherms

The polymerisation process may well produce heat. This is a problem for those manufacturers producing the raw plastics (normally in pellet form for thermoplastics), but there is a high degree of control of the processes involved, and it rarely causes a problem. However, the cross-linking or curing process involved in the manufacture of thermosetting materials may also be exothermic; for example, if blocks of polyurethane foam are stored before the exothermic curing process is complete, self-heating may occur, leading to spontaneous combustion (Section 7.7.2).

9.2.5 Catalysts

Various types of catalyst are used in polymerisation processes including acids, alkalis, complex organo-metallic compounds and organic peroxides.

Acids and alkalis present well known hazards. An organo-metallic catalyst may be in the form of a slurry in flammable solvents: some of these compounds, such as aluminium triethyl, react violently with water. Organic peroxides are oxidising agents and, therefore, present a considerable fire risk. **Under some conditions such materials can be explosive.**

9.2.6 Flammable solvents

Flammable liquids, such as acetone, methyl ethyl ketone, toluene, industrial alcohol and methyl alcohol are widely used as solvents in various processes and also as cleaning fluids, so may be present in fires on industrial premises. See also Chapter 8.

9.2.7 Dusts

Some processes produce fine plastic dusts. **These may present an explosion hazard if dispersed as a suspension in air.**

9.2.8 Self-extinguishing plastics

Many plastics are described as "self-extinguishing". While PVC and phenol-formaldehyde resins are naturally so, others may be made self-extinguishing by chemical changes in the polymer molecule or by the use of special additives.

The term 'self-extinguishing' means that, while a flame may (or may not) cause the material to burn, it will not continue to burn if the applied flame is removed. However, the term refers to the performance of plastics in a specific small scale test. In a fire situation, it will burn if surrounded by other burning materials, or perhaps on its own if a large enough area of the plastic has been ignited to produce flames which are self-sustaining.

> **'self-extinguishing' does not mean 'non-flammable' or 'non-combustible'.**

9.3 Monomer hazards

Monomers by definition are reactive compounds capable of polymerisation. Some, like ethylene, do not polymerise very easily and need exactly the right conditions of temperature and pressure, perhaps with a catalyst. Others, like styrene, may polymerise by accident, due to the presence of impurities, water, heat or other causes, and when this happens a great deal of heat may be given out. Some monomers have to be transported with a polymerisation inhibitor added to prevent the process occurring spontaneously.

As these monomers are mostly poor conductors of heat, the heat cannot get away easily, temperatures may rise and a fire may result. In addition to this problem, **monomers are flammable, and some are toxic.** It is also possible that some monomers in bulk quantities could start to polymerise in a fire with the ensuing added hazard of the heat given out by the polymerisation reaction.

Some of the more notable monomers used in the plastics industry are detailed below, but the number of monomers is very large and ever increasing, so the list cannot be considered as exhaustive.

Acrylonitrile

- colourless, partially water-soluble, flammable liquid with faint, pungent smell;

- polymerises explosively with some organic peroxides or concentrated caustic alkalis;

- highly toxic: can be absorbed through the skin and also through leather; and

- used as a starting material in the manufacture of ABS (acrylonitrile-butadiene-styrene) plastics and certain synthetic rubbers.

Butadiene

- vapour at room temperature and pressures, but easily liquefiable at room temperature;

- polymerises readily, especially in the presence of peroxide catalysts or air;

- flammable; and

- slightly toxic, narcotic in high concentration.

Epichlorhydrin

- used in the manufacture of epoxy resins;

- colourless, slightly water-soluble liquid with an irritating odour;

- polymerises exothermally with acids, bases and some salts; and

- highly toxic material and in fires may produce toxic gases including phosgene.

Methyl methacrylate

- clear liquid with acrid odour, used in the manufacture of acrylics (poly-(methyl-methacrylate));
- flammable;
- toxic;
- polymerises exothermally with peroxide catalysts; and
- normally stabilised, but heat accelerates the polymerisation.

Styrene

- used in the manufacture of polystyrene plastics and fibre-glass polyester resins;
- slightly yellow liquid, strong smell;
- normally stable, polymerisation greatly accelerated by heat or added peroxides;
- exothermic polymerisation: risk of fire and even explosion; and
- moderately toxic, vapour is an irritant to the eyes.

Vinyl acetate

- colourless, slightly water-soluble, flammable liquid, faint odour;
- polymerises with organic peroxides or when heated; and
- low toxicity, may act as an eye irritant.

Vinyl chloride

- sweet-smelling vapour at room temperature and pressure, easily liquefied;
- severe explosion hazard when exposed to heat or flame;
- moderately toxic acts as an anaesthetic in high concentrations;
- liquid may cause freeze burns due to rapid evaporation;
- flammable; and
- combustion gases contain hydrogen chloride which is both toxic and corrosive.

9.3.1 Intermediates and hardeners

Isocyanates

- used as intermediates in polyurethane foam manufacture;
- mostly brown liquids, slightly water-soluble, characteristic odour;
- skin irritants, may cause dermatitis, toxic by skin absorption;
- flammable – emit toxic gases when on fire;
- isocyanate vapours cause bronchial spasm repeated exposure may bring sensitisation;
- great caution should be exercised in dealing with them; and
- made harmless using special solutions of ammonia in water, to which an emulsifying agent has been added.

> In manufacture dangerous exothermic reaction causing fires can readily occur where toluene di-isocyanate (TDI-) based foams are produced, but is much less likely where the more extensively used diphenyl methane di-isocyanate (MDI) foams are made.

Chlorosilanes

- intermediates in silicone plastic manufacture;
- mostly fuming clear liquids;
- highly toxic;

Physics and Chemistry for Firefighters 73

- flammable; and

- react with water to produce hydrogen chloride gas – reaction is strongly exothermic in many cases.

Epoxides

- **Amine** hardeners – generally toxic. Some may cause dermatitis.

Physics and Chemistry for Firefighters

Chapter 10

Chapter 10 – Other Combustible Solids

10.1 Wood

Wood is a complex polymeric material of natural origin. In spite of the widespread use of synthetic materials, wood still accounts for a high proportion of the combustible material which is used in buildings, not only as fittings and furniture, but also as structural members. The principal constituent of wood is cellulose, a polymer of D-glucose, which occurs in all higher plants.

There is a high water content in wood and the difference in moisture content between green and well-dried wood is significant in regard to fire risk. Considerable quantities of heat are required to dry timber, due to the high latent heat of vaporisation of water.

When wood is heated, decomposition starts at temperatures of around 170°C, forming char, with the evolution of carbon dioxide, carbon monoxide and water. The proportion of flammable vapours released at this stage is low. Above 300°C, the decomposition process which produces flammable vapours becomes the dominant pyrolysis reaction, but they will be mixed with some CO_2 and H_2O vapour from the char-forming process which still occurs, but no longer dominates. This mixture of gases and vapours is less flammable than the decomposition products from, for example, polyethylene which will be 100% hydrocarbon, undiluted by non-flammable gases. This, and the fact that there is always a significant amount of char produced which provides protection to the wood underneath, accounts for the remarkable fire properties of wood. For example, (i) thick sections of wood cannot burn in isolation and (ii) thick timber beams can survive longer in a fire than unprotected steel beams, because the char forms a protective layer around the sound timber below.

Many methods are available to reduce the combustibility of wood. The most successful involve impregnating the wood with chemicals (e.g., ammonium phosphates, etc.) which catalyse the char-forming reaction at the expense of the decomposition process which produces flammable vapours. The resulting vapours are of very low flammability (mainly a mixture of CO, CO_2 and H_2O) and will not support flame, or contribute significantly to a fire when other materials are burning.

10.2 Coal

Coal is a very complex mixture of carbon and a variety of resinous organic compounds. There are many varieties, with harder coal containing more carbon. Plates of inorganic, noncombustible materials are also found in coal: these consist of limestone and compounds of iron, magnesium and manganese.

In large heaps, such as those used for storage of coal at power stations, self-heating can occur, which may lead to spontaneous combustion (see section 7.7). The smaller the coal particles, the the greater the danger. It is promoted by moisture, and the greater the oxygen content of the coal, the greater the danger, so that coal, especially pulverised coal containing more than 10 per cent of oxygen, may be dangerous in storage. In addition, **coal dust can form an explosive mixture in air**.

Storage heaps must be kept as free as possible from excess of air and protected from external sources of heat. Coal is sometimes sprayed with a high flash-point mineral oil that reduces dustiness and protects coal surfaces against oxidation. Fires in coal stacks are dealt with in the Manual of Firemanship, Part 6c.

10.3 Metals

Three-quarters of all the elements are metals. To a chemist a metal is a substance which can lose electrons and form positive ions (ions are charged atoms or groups of atoms – see Chapter 1). In addition, metals tend to have a group of properties associated with them; if an element possesses most of them, we describe it as a metal.

10.3.1 Properties of metals

Metals will show most of the properties listed here, although there are exceptions to most of these. For example, most metals are malleable and ductile - can be hammered into shape and can be drawn out into wire – but antimony is very brittle and will shatter if hammered!

(i) all, except mercury, are solids at room temperature, though they have a wide range of melting and boiling points;

(ii) they form positive ions;

(iii) they are malleable and ductile;

(iv) they are good conductors of heat and electricity;

(v) they can form alloys;

(vi) the oxides and hydroxides are basic (and so can be alkali solutions in water); and

(vii) most dissolve in mineral acids, normally releasing hydrogen.

Metals show a wide range of chemical properties, and range from dangerously reactive metals such as sodium and potassium, to inert metals such as platinum and gold. Metals can be arranged in an 'Activity Series' (Table 10.1).

In this Table, the most reactive metals are at the top and the least reactive at the bottom. Whatever chemical property is considered, those metals at the top of the series react most vigorously, indeed often violently, and those at the bottom react slowly or not at all.

Table 10.1 *The activity series of metals*

Metal		Occurrence	Reaction with water
Potassium	K	Never found	React with
Sodium	Na	uncombined	cold water
* Barium	Ba	with other	to yield
Strontium	Sr	elements	hydrogen
Calcium	Ca		
Magnesium	Mg	Rarely found	Hot metals
Aluminium	Al	uncombined	decompose
* Chromium	Cr	with other	water and
Manganese	Mn	elements	Burning metals
Zinc	Zn		decompose
* Cadmium	Cd		steam
Iron	Fe		
Cobalt	Co		Very little
Nickel	Ni		reaction unless
Tin	Sn		at white heat
* Lead	Pb		
† HYDROGEN	H		
*Bismuth	Bi	Sometimes	Inactive with
Copper	Cu	found	water or steam
* Mercury	Hg	uncombined	
Silver	Ag	with other	
		elements	
Platinum	Pt	Found	
Gold	Au	uncombined	
		with other	
		elements	

* Indicates that breathing apparatus must be worn in an incident involving these metals.
† Although not a metal, hydrogen is included as it also forms a positive ion

Although hydrogen is not a metal, it is included in the Table as it also forms a positive ion. Many important metal reactions involve displacement of hydrogen either from water or from acids.

10.3.2 Reaction of metals with water or steam

It is obviously important for the firefighter to understand how these heated metals will react with water as, usually, water will be the most readily available firefighting medium. It will be shown that for some metals, though, the addition of water could be dangerous or disastrous.

- **Potassium to calcium**

These metals react immediately with water to release flammable hydrogen gas and leave a metal hydroxide. In some cases the hydroxide so formed is itself a corrosive alkali. In the case of potassium the reaction is so vigorous that the metal seems to ignite immediately on contact with water. A small piece of sodium will move rapidly over the surface of water and if prevented from doing so, will ignite. Larger pieces of these metals are in danger of explosion on contact with water. Calcium reacts steadily with cold water but vigorously with hot.

- **Magnesium to iron**

These metals react little with cold water, even when powdered. At higher temperatures, the reaction rate increases and a steady flow of hydrogen is produced by reaction with steam. If the metals are already burning the reaction with cold water becomes very fast, producing a lot of hydrogen which may lead to an explosion. Going down the series, the rate of reaction decreases until, with iron, there is little reaction unless the red-hot metal is exposed to steam.

- **Cobalt to lead**

Here the white-hot metals must be treated with steam before reaction will take place.

- **Bismuth to gold**

These metals do not react with water or steam as they are below hydrogen in the Activity Series.

10.3.3 Reaction with oxygen

Metals at the top of the Activity Series react most readily in air*. Sodium and potassium are so reactive that they are stored in paraffin oil to prevent oxygen reaching them. Many other metals will burn in air or oxygen with increasing difficulty going down the series. Even metals like tin and lead will burn at very high temperatures.

---

* Note that reaction in pure oxygen is always much more vigorous than in air.

When a metal is powdered, it presents a very large surface area compared with a block of the same mass. Combustion is made much easier as the powder or dust particles are so small and can be heated extremely rapidly to the temperature at which they will burn. **Some metallic dusts can burn or explode spontaneously when dispersed in air.** When this occurs at ordinary temperatures the material is said to be pyrophoric (e.g., "Raney Nickel", which is a finely divided form of Nickel, used as a catalyst).

Many flammable metal powders and dusts are pyrophoric, especially magnesium, calcium, sodium, potassium, zirconium, hafnium. Some metal powders will burn in carbon dioxide and nitrogen (e.g., magnesium) or under water. Metal powders when damp may also cause fires and explosions, even in the absence of air and often without warning, and in the absence of heat.

10.4 Sulphur

This is usually found either as a yellow powder (known as 'flowers of sulphur') or as yellow crystals, but it is sometimes produced as blocks or sticks. It burns with a blue flame to give sulphur dioxide:

$$S + O_2 = SO_2$$

Sulphur is used in the manufacture of rubber, in sulphur compounds, such as sulphur dioxide and sulphuric acid and in certain drugs. **It has a low toxicity, but the dust presents an explosion hazard.** Sulphur dioxide, however, is a highly toxic gas with a sharp pungent odour which can be easily liquefied under pressure at ambient temperatures. It has many uses, especially as a bleaching agent and as a food preservative.

Hydrogen sulphide (H_2S), which is also known as sulphuretted hydrogen, is formed as a by-product from many chemical processes, including the decomposition of organic sulphur compounds; for this reason it is frequently found in sewer gases. Hydrogen sulphide has a characteristic odour of rotten eggs and is highly toxic. **It is flammable and under certain conditions can produce an explosion risk.**

10.5 Phosphorus

The element phosphorus is extremely reactive and is found in nature combined with other elements, mostly as phosphates (compounds containing the PO_4 group). It is also present in all living matter. The pure element exists in two different forms: red phosphorus and white (or yellow) phosphorus. Their properties are itemised in the Manual of Firemanship, Part 6C, Section 16. White phosphorus is extremely dangerous as it will ignite in air at temperatures as low as 30°C giving dense white clouds of toxic fumes of phosphorus pentoxide:

$$4P + 5O_2 \rightarrow 2P_2O_5$$

White phosphorus should never be touched with the bare hands as their warmth may cause ignition; moreover, phosphorus burns heal very slowly.

Red phosphorus is relatively safe if handled with care, and is used in making safety matches. The white form, because of its toxicity, is converted to phosphorus sulphide (P_4S_3) for use in non-safety matches.

Inorganic phosphates are crystalline solids which are normally safe unless one of the toxic metals is involved. Some are used as fertilisers.

Organic phosphates can be very toxic: some are used as pesticides and, for some of these, a few drops on the skin can prove fatal.

Physics and Chemistry for Firefighters

Chapter 11

Chapter 11 – Extinguishing Fires

> The ways in which burning can be stopped, and so the mechanisms which act when a fire is put out, have been dealt with in the previous chapters.
>
> In this chapter, the various media used to put out different types of fire are examined.
>
> The 'weapon' used against any given type of fire will depend upon the nature of the materials involved and the size and intensity of the fire.

11.1 Classification of fires by type

The current British/European Standard BS EN 2: 1992 *Classification of fires* defines four categories of fire, according to the type of material burning.

Class A
These are fires on solid materials, usually organic, leaving glowing embers. Class 'A' fires are the most common and the most effective extinguishing agent is generally water in the form of a jet or spray.

Class B
These are fires involving liquids or liquefiable solids. For the purpose of choosing effective extinguishing agents, flammable liquids may be divided into two groups: those that mix (are miscible) with water and those that do not (are immiscible).

Extinguishing agents are chosen according to whether the liquid fuel will mix with water or not. Agents which may be used include water spray, foam, light water, vaporising liquids, carbon dioxide and dry chemical powders.

Class C
These are fires involving gases or liquefied gases in the form of a liquid spillage, or a liquid or gas leak, and these include methane, propane, butane, etc. Foam or dry chemical powder can be used to control fires involving shallow liquid spills, though water in the form of spray is generally used to cool the containers.

Class D
These are fires involving metals. Extinguishing agents containing water are ineffective, and even dangerous. Carbon dioxide or dry chemical powders containing bicarbonate will also be hazardous if applied to most metal fires. Powdered graphite, powdered talc, soda ash, limestone and dry sand are normally suitable for Class D fires. Special fusible powders have been developed for fires involving some metals, especially the radioactive ones.

Electrical fires
Electrical fires are not treated as a class of their own, since any fire involving, or started by, electrical equipment must, in fact, fall into one of the other categories.

The normal procedure for dealing with an electrical fire is to cut off the electricity and use an extinguishing method appropriate to what is burning.

If this cannot be done with certainty, special extinguishing agents will be required which are nonconductors of electricity and non-damaging to equipment. These include vaporising liquids, dry powders and carbon dioxide. Very fine water mists have proven to be very effective at extinguishing fires using very little water, and their development has been hastened recently as they are seen to be an environmentally friendly replacement for halons. The rapid cooling that can be brought about by carbon dioxide extinguishers may affect sensitive electronic equipment – though it is the smothering effect of the gas, rather than the cooling which extinguishes the fire.

11.2 Classification of fires by size

To describe the size of a fire, the Central Fire Brigades Advisory Council has made the following recommendation:

Major fire	20–jets (or more)
Large fire	8–19 jets
Medium fire	3–7 jets
Small fire	1–2 jets, or 3 + hose reels
Minor fire	1–2 hose reels, or hand extinguishers.

11.3 Extinguishing fire: Starvation, smothering, cooling

We have seen from the *triangle of combustion* (see Figure 7.1), that three things are needed to allow burning to take place:

- a combustible material – fuel,
- oxygen, usually from the air, and
- enough heat to bring the material to a certain minimum temperature.

Fire extinction is largely a matter of depriving the fire of one or more of these factors, so methods of extinguishing fire can be classified in terms of removing these factors:

- Starvation – limiting fuel.
- Smothering – limiting oxygen.
- Cooling – limiting temperature.

In practice, fire extinguishing methods often use more than one of these principles, but it will be convenient to group them according to the main principle involved.

11.3.1 Starvation

Fires can be starved of fuel (Figure 11.1, top) in three ways:

1. By removing potential fuel from the neighbourhood of the fire. For example, by:

- draining fuel from burning oil tanks;
- working out cargo at a ship fire;
- cutting trenches or creating fire breaks in, for example, peat, heath and forest fires, demolishing buildings to create a fire stop; and
- counter-burning in forest fires.

2. By removing the fire from the mass of combustible material – for instance, pulling apart a burning haystack or thatched roof.

3. By dividing the burning material into smaller fires which may be left to burn out or which can be extinguished more easily by other means. The beating out of a heath fire owes much of its effectiveness to this.

Figure 11.1 The triangle of combustion.
Top: starvation – or the limitation of the combustible material. Centre: smothering – or the limitation of oxygen. Bottom: cooling – or the limitation of temperature.

11.3.2 Smothering

If the oxygen supply to the burning material can be sufficiently reduced, burning will cease (Figure 11.1, centre).

The general procedure in methods of this type is to try to prevent fresh air from reaching the to the seat of the fire and so to allow the combustion to reduce the oxygen content in the confined atmosphere until it extinguishes itself. This is less effective where, as in the case of celluloid, the burning material contains within itself in a chemically combined form the oxygen it requires for combustion.

Smothering is the principle behind snuffing out candles and capping oil well fires. The battening down of a ship's hold when a fire breaks out below decks will sometimes hold the flames in check until port is reached. Small fires, such as those involving a person's clothing, can be smothered with a rug, blanket, etc., while the use of sand or earth on a small metal fire is a further instance of the same principle.

Foam is an important practical smothering agent. Foams form a "blanket" over the burning surface and so separate the fuel from the air, thus preventing fuel vapours from mixing with air while at the same time shielding the surface from direct heat transfer from the flames.

Fires can be smothered with a cloud of fine dry powder, usually sodium bicarbonate, shot from a pressurised extinguisher, though research suggests that chemical interaction (inhibition) by the powder may be as important as the smothering action.

Another technique using the smothering principle is the use of **ternary eutectic chloride** powder for use on metal fires. This is applied using a gas cartridge pressurised extinguisher. The fusing temperature of the powder is in the region of 580°C, and it is applied to form a crust over the burning metal depriving it of oxygen from the air.

Inert gases such as nitrogen and carbon dioxide can be used to smother a flame temporarily. If these gases are vigorously discharged in the immediate vicinity of the fire, the oxygen content of the atmosphere may be reduced to such an extent that burning cannot be supported. Patented mixtures of inerting gases are now extensively used instead of halons in computer installations. "Total flooding systems" are used to protect special risks such as computer installations and rare book collections in libraries. This requires that the inerting gas is released into a closed space, as the appropriate inerting concentration must be reached following the discharge, and then maintained.

For larger fires, however, inerting agents aren't so useful, as the convection currents set up are sufficiently powerful to dilute the inert blanket before the extinguishing action can take effect. Strong winds may have the same effect. Applying inerting agents in a liquid form, which is then vaporised by the fire likely to be more effective, particularly as the burning region is also cooled by this. However, inerting gases can be used to great effect in enclosed environments such as electrical cabinets.

Very fine water mists have been shown to be able to extinguish fires using very small amounts of water. These have shown their worth in situations where halons would previously have been used and on offshore installations. They act mainly by inerting: a great deal of water vapour is created when water mist is discharged into a confined space with hot surfaces, and this smothers the fire. As the droplets are so small, they evaporate very quickly, and can rapidly smother the flames. Their small size which makes them such a good extinguishing agent also means, unfortunately, that they are easily swept away from the fire by opposing air movement and so are not suitable for fighting larger fires in unconfined spaces.

In the 1970's and 80's, halogenated hydrocarbons or **halons** were developed and used extensively as extinguishants. The first and probably the simplest of these was **carbon tetrachloride**, but it is toxic and its use was soon discontinued. A number of others of lesser toxicity were developed and found favour.

Many halons have been considered. They are identified by numbers which denote how many carbon and halogen atoms are in the molecule. The first digit gives the number of carbon atoms, the second gives the number of fluorine atoms, the third gives the number of chlorine atoms and the fourth gives the number of bromine atoms. A fifth digit may, or may not, be present which gives the number of iodine atoms. For example, bromochlorodifluoromethane has a formula CF_2ClBr, and so is known as Halon 1211.

These vaporising liquids act partly as inerting blankets similar to those mentioned in the preceding section, but mainly by chemical interference (inhibition) with the chain reactions in the flame, "mopping up" free radicals.

Although extremely effective, halons are no longer used except under some exceptional circumstances, as they are known to be very harmful to the earth's protective ozone layer. No more will be manufactured, though such as are already in stock will continue to be used in fixed installations where there is perceived to be a special risk, such as in certain military situations.

Fires can also be extinguished by separating the fuel from the flame by blasting it away. This is what happens when a candle is blown out and, on a larger scale oil well fires can be extinguished by the blast from exploding dynamite. This method does not work simply by depriving the flame of fuel, but also by making the flame unstable when air is supplied at high velocity in the vicinity of the fuel surface.

11.3.3 Cooling

If the rate at which heat is generated by combustion is less than the rate at which it is lost from the burning material, burning will not continue. (Figure 11.1, bottom).

So, to extinguish a fire by cooling, the rate at which heat energy is lost from the burning material must be increased by removing some of the heat energy. This reduces the temperature of the burning mass, reducing the heat release rate. Eventually, the rate at which heat is lost from the fire may be greater than the rate of heat production and the fire will die away.

Cooling the fuel is the main way in which water is used to extinguish fires. There are many variations: for example, a tank fire involving a high flashpoint oil (boiling point >> 100°C) can be extinguished by a high velocity sprinkler spray which apparently produces a water-in-oil emulsion at the surface, thus causing rapid cooling. This neatly avoids the problem of water sinking to the bottom of the tank before it has had much effect on the temperature of the surface layer.

When it is applied to a fire, the extinguishing medium – water for example – itself undergoes changes as it absorbs heat from the fire:

(a) its temperature will rise;
(b) it may evaporate (boil);
(c) it may chemically decompose (not water); and
(d) it may react chemically with the burning material.

For the extinguishing medium to achieve maximum effect, it is clear that the quantity of heat energy absorbed when these changes occur should be as high as possible. That is to say that, referring to the points above in order, in a good coolant, the following properties should be as **high** as possible:

● the specific heat capacity;

● the latent heat of vaporisation; and

● the heat of decomposition.

The action of water depends predominantly on (a) and (b), the latter being far more important: it takes about six times as much heat to convert a given mass of water at its boiling point into steam as is required to raise the temperature of the same amount of water from the usual atmospheric value to its boiling point. Water is most efficiently used if it is applied to a fire in liquid form and in such a way that as much as possible is converted to steam. The smothering effect of the steam produced at the seat of the fire is thought to play a part in assisting in the extinguishing process.

In all fire-fighting operations where water is used it should be the aim to ensure that the proportion of water which escapes from the building in liquid form applied should be as low as possible.

When the heat of a fire is considerable, as in its early stages, the steam formed will not be visible, but as the temperature falls the steam will condense above the fire. This is widely recognised by experienced fire-fighters as a sign that a fire is being brought under control.

On the basis of thermal capacity and latent heat of vaporisation, water is an excellent fire extinguishant, since both figures are high. This fact, combined with its availability in large quantities, makes it by far the most useful fire extinguishant for general purposes. The role of decomposition is insignificant in the case of water, but certain substances, for example carbonates, absorb heat in this way (see the reference to dry powder extinguishers under Section 2, 'Smothering').

Water does not react with ordinary materials, but may prove dangerous with some fuels, evolving heat rather than absorbing it. Moreover, the reaction may result in the formation of a flammable product, thus adding fuel to the fire. The action of water on burning magnesium exemplifies both these effects, since it reacts with the metal exothermically (i.e., producing heat) with the formation of hydrogen, which is readily ignited. In the case of other media the reaction products may be undesirable in other senses, as in the case of the halons which can produce toxic gases which can be hazardous in enclosed spaces.

11.4 Fire extinguishing media

11.4.1 Water

Water is the most efficient, cheapest and most readily available medium for extinguishing fires of a general nature. It is used by fire brigades for the majority of fires, although the methods of application have undergone a number of improvements.

If more water is applied than is actually required to contain and extinguish the fire, the surplus will drain off and may seep through floors and perhaps cause more damage to goods and property than that caused by the fire itself. Accordingly, the method of applying water to a fire varies according to the size of the fire.

If only small quantities are required, portable water extinguishers or hand pumps may be sufficient. Hose reels are used for larger fires. These are fed from a tank on the appliance and water is pumped through the tubing on the reels by means of a built-in pump. For major fires, greater quantities of water are necessary: the built-in pumps driven by the vehicles' engines are often capable of pumping up to 4500 litres per minute, giving the water the necessary energy to provide adequate throw to penetrate deep into a building.

A variation in the application of water can be made using nozzles that produce jets or sprays ranging from large size droplets down to "atomised" fog. Judicious use of this type of application can not only cut down the amount of water used, minimising water damage, but also ensure that it is used to greatest effect. Atomised spray (fog) nozzles have become standard equipment on fire brigade appliances in the UK. They are quite effective when used in the correct situations, but their range is limited. Special pumps and ancillary equipment are used with high pressure fog, giving a greater range of application.

11.4.2 "Inert gas"

On cargo ships, a fire in a hold may be contained by "inerting" the space using the exhaust gases from the ship's engines to displace air. These gases have low oxygen and high carbon dioxide concentrations. (In the past, steam has been tried as a fire suppression agent to control fires in the petrochemical industry, but it is very expensive, requiring a fixed installation and an available source of high pressure steam.)

11.4.3 Foam

Firefighting foams have been developed primarily to deal with the hazards posed by liquid fuel fires.

Although water is used for most firefighting incidents, it is generally ineffective against fires involving liquid fuels. This is because water has a density that is greater than most flammable liquids so, when applied, it quickly sinks below their surfaces, often without having any significant effect on the fire.

Finished firefighting foams, on the other hand, consist of bubbles that are produced from a combination of a solution of firefighting foam concentrate and water that has then been mixed with air. These air filled bubbles form a blanket that floats on the surface of flammable liquids. In so doing, the foam blankets help to knock down and extinguish these fires in the following ways:

- by excluding air (oxygen) from the fuel surface;

- by separating the flames from the fuel surface;

- by restricting the release of flammable vapour fro m the surface of the fuel;

- by forming a radiant heat barrier which can help to reduce heat feedback from flames to the fuel and, hence, reduce the production of flammable vapour; and

- by cooling the fuel surface and any metal surfaces as the foam solution drains out of the foam blanket. This process also produces steam which dilutes the oxygen around the fire;

The main properties of firefighting foams include:

- Expansion: the amount of finished foam produced from a foam solution when it is passed through foam-making equipment;

- Stability: the ability of the finished foam to retain its liquid content and to maintain the number, size and shape of its bubbles. In other words, its ability to remain intact;

- Fluidity: the ability of the finished foam to be projected on to, and to flow across, the liquid to be extinguished and/or protected;

- Contamination resistance: the ability of the finished foam to resist contamination by the liquid to which it is applied;

- Sealing and resealing: the ability of the foam blanket to reseal should breaks occur, and its ability to seal against hot and irregular shaped objects;

- Knockdown and extinction: the ability of the finished foam to control and extinguish fires; and

- Burn-back resistance: the ability of the finished foam, once formed on the fuel, to stay intact when subjected to heat and/or flame.

The amount of air added to the foam solution depends on the type of equipment used. Hand-held foam-making branches generally only mix relatively small amounts of air into the foam solution. Consequently, these produce finished foam with low expansion (LX) ratios, that is to say, the ratio of the volume of the finished foam produced by the nozzle, to the volume of the foam solution used to produce it, is 20:1 or less. Other equipment is available which can produce medium expansion foam (MX) with expansion ratios of more than 20:1 but less than 200:1, and high expansion foam (HX) with expansion ratios of more than 200:1 and possibly in excess of 2000:1.

There are a number of different types of foam concentrate available. Each type normally falls into one of the two main foam concentrate groups, that is to say, they are either protein based or synthetic based, depending on the chemicals used to produce them.

Protein based foam concentrates include:

Protein (P);
Fluoroprotein (FP);
Film-forming fluoroprotein (FFFP); and
Alcohol resistant FFFP (FFFP-AR).

Synthetic based foam concentrates include:

Synthetic detergent (SYNDET);
Aqueous film-forming foam (AFFF); and
Alcohol resistant AFFF (AFFF-AR).

The characteristics of each of these foam concentrates, and the finished foams produced from them, vary. As a result, each of them has particular properties that makes them suitable for some applications and unsuitable for others.

Various types of surface active agents (or surfactants) are added to many firefighting foam concentrates. These are used to reduce the amount of fuel picked up by the finished foam on impact with fuel and to increase the fluidity of the finished foam. Surface active agents are also used as foaming agents because they readily produce foam bubbles when mixed with water.

In film-forming foam concentrates, surface active agents form an aqueous film of foam solution which, in certain conditions, can rapidly spread over the surface of **some** burning hydrocarbons to aid knockdown and extinction. This ability can make them ideal for use in certain firefighting situations such as aircraft crash rescue. However, the associated foam blanket tends to collapse quickly, so providing very poor security and resistance to burnback.

Water-miscible liquids, such as some polar solvents, mix freely with water and can quickly attack finished foams by extracting the water they contain. This rapidly leads to the complete destruction of the foam blanket. Fires involving these liquids can be extinguished by diluting them with large quantities of water. However, containment of the resulting mixture can cause problems and the application of sufficient quantities of water to achieve extinction can take a long time. Consequently, 'alcohol resistant' foam concentrates have been developed to deal with these particular types of liquid.

Further technical detail regarding foam will be found in the Fire Service Manual – Volume 1 – 'Fire Service Technology, Equipment and Extinguishing Media – Firefighting Foam' and details of operational use – in Volume 2 – Fire Service Operations – 'Firefighting Foam'.

11.4.4 Vapourising liquids

This category consists mainly of halons as discussed in Section 11.3.2. Halons have the property of vapourising rapidly when released from their pressurised container. The vapours are heavier than air, but when entrained into the flames, they inhibit the chain reactions and suppress flaming.

Due to environmental concerns, halons have largely been replaced with inerting gases (see Section 11.4.5) and fine water mists.

11.4.5 Carbon dioxide and inert gases

At normal temperatures, carbon dioxide is a gas 1.5 times as dense as air. It is easily liquefied and bottled in a cylinder, where it is contained under a pressure of approximately 51 bars at normal temperatures. When discharged from the cylinder, cold CO_2 vapour and some solid CO_2 are expelled from the "horn", which rapidly cools in the process. The solid quickly sublimes, and some of the liquid CO_2 evaporates to maintain the pressure in the cylinder. The gas, however, extinguishes by smothering, effectively reducing the oxygen content of the air. About 20 to 30 per cent is necessary to cause complete extinction, depending on the nature of the burning material. In fact, materials which have their own oxygen "supply" will continue to burn, as will any material that tends to react with the carbon dioxide, such as burning magnesium. Apart from these considerations, carbon dioxide is quick and clean, electrically non-conducting, non-toxic and non-corrosive. Most fabrics are unharmed by it.

For special risk situations, such as in transformer rooms and rare book collections in libraries, total flooding of the compartment may be required. For this, fixed carbon dioxide installations may be built in. However, although it is non-toxic, it is an asphyxiant at the concentrations necessary to extinguish a fire. The operation of total flooding CO_2 systems requires prior evacuation of all personnel.

Carbon dioxide is also available in bulk to fire authorities, by special arrangement with certain manufacturers who have agreed to supply tankers containing 10 tonnes of liquid to any fire on request.

11.4.6 Dry chemical powders

New problems have been produced for the firefighter by the use in industry of an ever widening range of risks and materials.

The rise of new plastics is one example of this, and the fabrication of reactive metals such as titanium, zirconium and beryllium is another. Sometimes, water cannot be used; on most fires involving burning metals, the result of applying water can be disastrous, often leading to an explosion. New methods of extinction have had to be evolved.

Chief among these are the dry chemical powders which are stored in cylinders under pressure, or which can be ejected by the release of gas under pressure.

The basis of most of these is sodium bicarbonate, which, with the addition of a metallic stearate as a waterproofing agent, is widely used as an extinguishant both in portable extinguishers and for general application in large quantities. Apart from stearates, other additives are sometimes used to decrease the bulk density and to reduce packing in the cylinder. Dry powder is very effective at extinguishing flame ("rapid knock-down"), and is particularly valuable in tackling a fire involving an incident in which someone's clothes have been soaked in flammable liquid, and ignited.

Dry chemical is expelled from containers by gas pressure and directed at the fire in a concentrated cloud by means of specially designed nozzles. This cloud also screens the operator from the flames and enables a relatively close attack to be made. Dry chemical powder can also be supplied in polythene bags for metal fires, as it is more effective to bury the fire under a pile of bags which melt and allow the contents to smother the fire.

Dry chemical powders are also tested for their compatibility with foam, as it was discovered that the early powders tended to break down foam. The two can complement each other on fires where foam is the standard extinguishant.

Ternary eutectic chloride powders have been developed for some metal fires, especially for the radioactive metals such as uranium and plutonium. These contain an ingredient which melts, then flows a little and forms a crust over the burning metal, effectively sealing it from the surrounding atmosphere and isolating the fire.

Some burning materials, such as metals, which cannot be extinguished by the use of water, may be dealt with by means of dry earth, dry sand, powdered graphite, powdered talc, soda ash or limestone, all of which act as a smothering agent.

Dry sand may also be used to prevent burning liquids, including paints and oils, from flowing down drains, basement lights, etc., and for confining shallow layers of such liquids, thus permitting the use of foam or spray branches. On no account should sand be used for extinguishing fires in machinery, such as electric motors, since its use may well necessitate dismantling the entire machine for cleaning, even though the fire damage is negligible.

11.4.7 Blanketing

Another fire extinguishing method is blanketing, which deprives the fire of oxygen. This is especially useful if someone's clothes are burning. The person should be laid down and covered or rolled in a rug, coat, jacket, woollen blanket, etc.

For dealing with fires in small utensils, such as those containing cooking fats, the best method is to smother the fire with a fire resisting blanket, or a cloth or doormat which has been wetted first.

11.4.8 Beating out

Small fires in materials, such as textiles, etc., may often be extinguished by beating them out, or by rolling and screwing up the burning material tightly to exclude the air. Beating is also the method normally employed to extinguish heath, crop and other similar fires in rural areas when water is not readily available.

APPENDIX A

Metrication

List of SI units for use in the fire service.

Quantity and basic or derived SI unit and symbol	Approved unit of measurement	Conversion factor
Length metre (m)	kilometre (km) metre (m) millimetre (mm)	1 mile = 1.609 km 1 yard = 0.914m 1 foot = 0.305m 1 inch = 25.4 mm
Area square metre (m²)	square kilometre (km²) square metre (m²) square millimetre (mm²)	1 mile² = 2.590 km² 1 yard² = 0.836 m² 1 foot² = 0.093m² 1 inch² = 645.2 mm²
Volume cubic metre (m³)	cubic metre (m³) litre (l) (= 10^{-3}m³)	1 cubic foot = 0.028 mJ 1 gallon = 4.546 litres
Volume, flow cubic metre per second (m³/s)	cubic metre per second (m³/s) litres per minute (l/min = 10^{-3}m³/min)	1 foot³/s = 0.028 m³/s 1 gall/min = 4.546 l/min
Mass kilogram (kg)	kilogram (kg) tonne (t) (1 tonne = 10^3kg)	1 lb = 454 kg 1 ton = 1.016 t
Velocity metre per second (m/s)	metre/second (m/s) International knot (kn) kilometre/hour (km/h)	1 foot/second = 0.305 m/s 1 Int. knot = 1.852 km/h 1 UK knot = 1.853 km/h 1 mile/hour = 1.61 km/h
Acceleration metre per second² (m/s²)	metre/second²	1 foot/second² = 0.305 m/s² 'g' = 9.81 m/s²
Force Newton (N)	kiloNewton (kN) Newton (N)	1 ton force = 9.964 kN 1 lb force = 4.448 N

Quantity and basic or derived SI unit and symbol	Approved unit of measurement	Conversion factor
Energy, work Joule (J) (= 1 Nm)	joule (J) kilojoule (kJ) kilowatt-hour (kWh)	1 British thermal unit = 1.055 kJ 1 foot lb force = 1.356 J
Power watt (W) (= 1 J/s = 1 Nm/s)	kilowatt (kW) watt (W)	1 horsepower = 0.746 kW 1 foot lb force/second = 1.356W
Pressure newton/metre^2 (N/m^2)	bar= 105 N/m^2 millibar (m bar) (= 10^2 N/m^2) metrehead	1 atmosphere = 101.325 kN/m^2 = 1.013 bar 1 lb force/in^3 = 6894 76 N/m^2 = 0.069 bar 1 inch Hg = 33.86 m bar 1 metrehead = 0.0981 bar 1 foothead = 0.305 metrehead
Heat, quantity of heat Joule (J)	joule (J) kilojoule (kJ)	1 British thermal unit = 1.055 kJ
Heat flow rate watt (W)	watt (W) kilowatt (kW)	1 British thermal unit/ hour = 0.293 W 1 British thermal unit/ second = 1.055 kW
Specific energy, calorific value, specific latent heat joule/kilogram (J/kg)	kilojoule/kilogram (kJ/kg) kilojoule/m^3 (kJ/m^3) joule/m^3 (J/m^3) megajoule/m^3 (MJ/m^3)	1 British thermal unit/ lb = 2.326 kJ/kg 1 British thermal unit/ft^3 = 37.26 kJ/m^3
Temperature degree Celsius (°C)	degree Celsius (°C)	1 degree centigrade = 1 degree Celsius

APPENDIX B

List of the elements with atomic number atomic weight and valency

Name of element	Symbol	Atomic number	Atomic weight	Valency
Actinium	Ac	89	227.0	3
Aluminium	Al	13	27.0	3
Americium	Am	95	243.0	3, 4, 5, 6
Antimony	Sb	51	122.0	3, 5
Argon	Ar	18	40.0	0
Arsenic	As	33	75.0	3, 5
Astatine	At	85	210.0	1, 3, 5, 7
Barium	Ba	56	137.0	2
Berkelium	Bk	97	249.0	3, 4
Beryllium	Be	4	9.0	2
Bismuth	Bi	83	209.0	3, 5
Boron	B	5	11.0	3
Bromine	Br	35	80.0	1
Cadmium	Cd	48	112.0	2
Calcium	Ca	20	40.0	2
Californium	Cf	98	251.0	
Carbon	C	6	12.0	2
Cerium	Ce	58	140.0	3, 4
Caesium	Cs	55	133.0	1
Chlorine	Cl	17	35.5	1
Chromium	Cr	24	52.0	2, 3, 6
Cobalt	Co	27	59.0	2, 3
Copper	Cu	29	63.5	1, 2
Curium	Cm	96	247.0	3
Dysprosium	Dy	66	162.5	3
Einsteinium	Es	99	254.0	
Erbium	Er	68	167.0	3
Europium	Eu	63	152.0	2, 3
Fermium	Fm	100	257.0	
Fluorine	F	9	19.0	1
Francium	Fr	87	223.0	1
Gadolinium	Gd	64	157.0	3
Gallium	Ga	31	70.0	2, 3
Germanium	Ge	32	73.0	4
Gold	Au	79	197.0	1, 3
Hafnium	Hf	72	178.5	4
Helium	He	2	4.0	0
Holmium	Ho	67	165.0	3
Hydrogen	H	1	1.0	1
Indium	In	49	115.0	3

Name of element	Symbol	Atomic number	Atomic weight	Valency
Iodine	I	53	127.0	1
Iridium	Ir	77	192.0	3, 4
Iron	Fe	26	56.0	2, 3
Krypton	Kr	36	84.0	0
Lanthanum	La	57	139.0	3
Lawrencium	Lw	103	257.0	
Lead	Pb	82	207.0	2, 4
Lithium	Li	3	7.0	1
Lutecium	Lu	71	175.0	3
Magnesium	Mg	12	24.0	2
Manganese	Mn	25	55.0	2, 3, 4, 6, 7
Mendelevium	Md	101	256.0	
Mercury	Hg	80	201.0	1, 2
Molybdenum	Mo	42	96.0	3, 4, 6
Neon	Ne	10	20.0	0
Neptunium	Np	93	237.0	4, 5, 6
Nickel	Ni	28	59.0	2, 3
Niobium	Nb	41	93.0	3, 5
Nitrogen	N	7	14.0	3, 5
Nobelium	No	102	253.0	
Osmium	Os	76	190.0	2, 3, 4, 8
Oxygen	O	8	16.0	2
Palladium	Pd	46	106.0	2, 4, 6
Phosphorus	P	15	31.0	3, 5
Platinum	Pt	78	195.0	2, 4
Plutonium	Pu	94	242.0	3, 4, 5, 6
Polonium	Po	84	210.0	2, 3, 4
Potassium	K	19	39.0	1
Praseodymium	Pr	59	141.0	3
Promethium	Pm	61	145.0	3
Protactinium	Pa	91	231.0	5
Radium	Ra	88	226.0	2
Radon	Rn	86	222.0	0
Rhenium	Re	75	186.0	2, 3, 4, 6, 7
Rhodium	Rh	45	103.0	3
Rubidium	Rb	37	85.5	1
Ruthenium	Ru	44	101.0	3, 4, 6, 8
Samarium	Sm	62	150.0	2, 3
Scandium	Sc	21	45.0	3
Selenium	Se	34	79.0	2, 4, 6
Silicon	Si	14	28.0	4
Silver	Ag	47	108.0	1

APPENDIX B continued

Name of element	Symbol	Atomic number	Atomic weight	Valency
Sodium	Na	11	23.0	1
Strontium	Sr	38	88.0	2
Sulphur	S	16	32.0	2, 4, 6
Tantalum	Ta	73	181.0	5
Technetium	Tc	43	99.0	6, 7
Tellurium	Te	52	128.0	2, 4, 6
Terbium	Tb	65	159.0	3
Thallium	Tl	81	204.0	1, 3
Thorium	Th	90	232.0	4
Thulium	Tm	69	169.0	3
Tin	Sn	50	119.0	2, 4
Titanium	Ti	22	48.0	3, 4
Tungsten	W	74	184.0	6
Uranium	U	92	238.0	4, 6
Vanadium	V	23	51.0	3, 5
Xenon	Xe	54	131.0	0
Ytterbium	Yb	70	173.0	2
Yttrium	Y	39	89.0	3
Zinc	Zn	30	65.0	2
Zirconium	Zr	40	91.0	4

Suggestions for further reading

Books about Combustion, Flame and Fire
J.F. Griffiths and J.A. Barnard "Flame and Combustion". Blackie Academic and Professional.

"Thermal Radiation Monograph" The Institution of Chemical Engineers.

References
D.D. Drysdale "An Introduction to Fire Dynamics". John Wiley 1985.

J.F. Griffiths, J.A. Barnard "Flame and Combustion" Blackie Academic and Professional 1995.

G.B. Grant D.D. Drysdale "A Review of the Extinction Mechanisms of Diffusion Flame Fires". Fire Research and Development Group Publication Home Office 6/96.

Books about Physics
Muncaster "A-Level Physics". Stanley Thornes (Publishers) Ltd.
Nelkon and Parker "Advanced Level Physics". Heinemann Educational Books

Books about Chemistry
Open University Foundation Course in Science

General
I. Asimov, "Asimov's New Guide to Science". (Penguin Books, 1985).

"The SFPE Handbook of Fire Protection Engineering" (second edition). Society of Fire Protection Engineers, Boston, Massachusetts, USA.

Acknowledgement
The author would like to thank the Editor of the *Fire Engineers Journal* for allowing portions of the article "Flames in Fires and Explosions" by J. R. Brenton and D.D. Drysdale to be reproduced herein.